The Long and The Short

A science guy explain the development of automotive technology... and much more!

AF797200

Steve Markman

How This Book Came To Be Written

I never was a car guy, and I didn't especially like to write, so how did I come to write a book about automotive technology? In the first article, I'll explain how I bought my 1950 MG TD roadster, so I won't get into it here. Actually, I'm still not really that much of a car guy, I just love my MG TD and have made many friends who own British sports cars. I'm really an airplane guy; I spent my career as an aerospace engineer, and hold a commercial pilot license.

Back in school, I hated writing, but mostly because it never was taught as a topic, and most of the writing we had to do was from English class, in which we had to discuss a story or poem we hated. Boring!! My interest in writing didn't develop until my freshman year in college. In the third and final English class I had to take, the instructor told us he was going to focus on writing, figuring we were sick of poetry, and feeling that good writing skills were more important than being able to discuss Shakespeare or Keats. He taught us how to organize a paper, construct an argument, and gave us about twenty key grammar elements to watch for that would make us grammatically correct about 99 percent of the time. We still had to read the required literature, but as long as we backed up our argument and used proper grammar in our papers, he gave us good grades.

By my junior year, I joined the staff of the engineering college magazine and wrote six articles that they published. They were fun to do, but still a lot of work writing by hand, even with the help of a typewriter. It wasn't until the 1980s, with the invention of the desktop computer and word processing software that I found writing to be fun. Maybe because now I could get my thoughts down quickly, almost as fast as they popped into my head and before I lost them. Then, I could fix the wording quickly so that the sentence was clear. Editing and rearranging thoughts so that the paper flowed nicely was *so* easy. In my professional career, I went on to write a dozen or more technical papers, and then four books, two about research aircraft, and two personal memoirs. Then I decided to call it quits; too much work for too little financial gain.

After joining the MG car club in Dayton, Ohio I wrote an article about my car. Promptly, I was asked to take over as editor of the club's newsletter. That became a monthly task, but it was fun. Sometimes I was short on material, so I'd research a topic of automotive interest to fill a few pages, explaining things as simply as I could. That turned into a few articles each year for the last fifteen or more years, and club members always thanked me for explaining technical things so clearly. Not every article was about automotive technology; sometimes they were just about something on my mind. Then, one day recently, I though "Hey, maybe I have enough here for another book." I guess some people never learn, so I hope you enjoy my collection of stories and learn about the many technologies that have evolved to make your car what it is today.

Table of Contents

My Car and My Club

My 1950 MG TD

My favorite picture, taken at the 2nd St Market in May 2004.

I've never been a car guy, so I really cannot tell you where, when, or why I fell in love with the MG TD, just that I always wanted one. By the late 1990s, I started thinking seriously, not for a real one, but a kit car. I never realized that real TDs were both available and affordable, so was mentally planning out a five-year project to build a replica. After months of internet searches, I finally found the Canadian company that made the fiberglass conversion kits. I also found dozens of adds for real TDs, and eventually shifted my thinking away from a replica to the real thing (BTW – I figured it'd cost me about $10,000 or more to build the kit version, but completed ones that were for sale seldom went for over $5,000).

I attended my first British Car Day in Dayton, Ohio in 2000. Upon finding the row of TD's, I asked one proud owner how I could buy one. He directed me to a distinguished gentleman in a straw hat standing next to his red TD, who he said was the president of the MG Club. This was how I met Phil Johnson. I repeated the question to Phil, who told me the best thing would be to join the club, since members usually know of any TDs that come on the market locally.

Doug Gillen hands the car keys to me in Sep 2000.

Within a few months, Phil got me in touch with former member Ernie Streifthau, who knew of a TD for sale in Franklin, a small town just south of Dayton. It'd been completely restored by former club member Doug Gillen. Doug had been an active member for many years, but his wife had had a hip replacement a few years earlier and now found it too difficult to get in and out of the TD. Doug continued driving to events by himself, but said it wasn't as much fun without his wife riding along, so decided to sell the car. He hadn't advertised it, but figured he'd part with it when the right person came along. I guess I was that person.

The car's original owner was a doctor in Atlanta. By the late 60s, he sold it to someone in Portsmouth, Ohio. It sat in a barn for the next twenty years. Doug bought it and spent about six years restoring it. (BTW, Doug gave me the name of the original owner, now long deceased. On a hunch one night, I looked up the name and found a doctor with the same name, but Junior, in Atlanta, possibly his son. I called and got a fax machine. I should try again one of these days.).

I drove the car home in September, 2000, only a few days short of what I eventually learned was its 50th birthday. I was too embarrassed to tell Doug that I never actually had driven a TD. He probably was shaking his head as I drove off, wondering if he'd made the right decision as he watched me trying to shift. The first thing that I learned was not to wear wide athletic shoes when driving a TD (you step on your own feet a lot). The next thing I learned while driving the car home was that a lot of nice-looking women you don't even know smile and wave (maybe they were thinking 'Doesn't the dumb s--t even know how to drive a stick?').

While Doug had to replace the wood frame and interior with modern reproductions, all the sheet metal was original. The engine and transmission also were original, although having been overhauled over the years, which is to be expected. The engine and body numbers match. One of the things I liked about Doug's restoration is that it was very authentic. The only changes he made from original were installing red, instead of factory-standard black carpet, adapting the engine to take a modern spin-on oil filter, replacing the vinyl on the dash with mahogany (although this was a common period upgrade), and painting the car red, instead of its original green (since he already owned a green TD).

Seat belts neatly placed for display. Note the temperature gauge on the bottom of the dash, just left of the steering wheel.

I've only made a few upgrades to the car since I've owned it. The first was to add seat belts, as I never felt comfortable without them. The belts have tan webbing to match the interior, and chrome latches. They look completely period-appropriate and I carefully cross them over the seats for car shows. Next came a temperature gauge and volt meter a few years later. I made a mounting bracket and placed it on the bottom edge of the dash where I could remove it easily for shows where the judges might know that it's not correct. The last upgrade came after I'd owned the car for ten years, when I replaced my Dunlop bias-ply tires with Kumho radials. The difference in ride and handling was incredible! The car feels like it is twice as heavy and twice as wide; it really hugs the road.

After having owned my TD for over twenty years, I love it as much as the day I first saw it. I do five or six car shows each summer. While I don't drive it great distances, I still love taking it out on hilly, winding country roads and just having fun.

How Many Owners?

My MG TD's original owner was a doctor who lived in Chamblee, Georgia, a suburb of Atlanta. He owned it for some years until he either sold or gave it to his nephew who lived in Marietta, Ohio (or was it Portsmith, as I said earlier…I don't remember). By then, the car was showing signs of age and was in need of a major rebuilding.

It remained in storage in this condition, never having been restored. In 1982, he sold the car to the gentleman in Franklin, Ohio, from whom I eventually purchased it. Doug performed a thorough disassembly and overhaul, this having been the second TD that he restored. He replaced most of the wood, recovered the seats and door panels, and overhauled the transmission. Doug overhauled existing items wherever possible, even leaving some older dents and damage (such as the patch on the oil pan where a piston rod had gone through it many years earlier). He paid immaculate attention to the authenticity of all details.

Throughout the restoration, Doug had a problem with obtaining title. It appears that the title still was in the name of the original owner, never having been transferred to the nephew in Marietta (or was it Portsmith) because one digit of the serial number on the original title did not match the actual serial number stamped on the car. Doug could not locate the nephew in Marietta, and attempted to contact the original owner, but by now he was living in a nursing home, was in mental decline, and had no recollection of the car.

Doug learned of an attorney in New York City who could fix title problems. For a fee of $125, he produced a New York State title that showed the attorney as the owner with a purchase price of $125. This title was signed over to Doug, who promptly obtained an Ohio title. Thus, depending on whether I count the nephew in Marietta and the attorney in New York, I'm either the third or the fifth owner. Either way, my TD has had a loving home since I bought it in 2000.

What a Club is All About

When I first started to think about buying an MG TD, the last thing on my mind was joining a club; I just wanted a car. As I explained earlier, I came to my first British Car Day in 2000, looked over the line of beautiful TDs on display (and other British cars, too), and asked one of the owners how I could find one. "Join the MG Car Club," and I was directed to a distinguished-looking gentleman sporting a goatee and wearing a straw hat. "He's our president, and the club can help you find a car." That's how I met Phil Johnson.

Needless to say, the rest is history. My first club meeting was an experience. I was the youngest person there by at least ten years. I also won the gumball rally (a $5 bill awarded to an attendee, something that didn't happen again for nearly fifteen years of being a member (could it have been fixed? Naaaaa…).

Soon after acquiring my TD, I asked Phil to talk me through doing an oil change and lubrication. He crawled all over and under my car with me (or maybe it was his car; I'm not so sure any more) and showed me everything to do, what grade oils and grease to use, trouble spots to keep an eye on, etc, etc.

Over the years, whenever I needed advice, help, or just a few words of support, I knew I always could count on club members. More than once, Steve Miller, owner of our local British car repair shop, passed by at a car show and crawled under the car with me for a quick look at something while people continued to walk past. Sometimes, I not only got advice, but someone to do the work. Several times, Jim Pesta has come up from Cincinnati to do work for me and other TD owners. Last year, he showed us how to replace the valve seals without pulling the head from the engine, using my car to demonstrate (no more cloud of smoke on startup), and then demonstrated some diagnostic techniques on another member's TD. Then, at British Car Day in Dayton, Jim conducted a diagnostic clinic, finding problems with several cars, including a worn distributor cap and rotor on my car (since replacing them, the car starts right up idles smoothly). Most recently, when I was disabled temporarily and couldn't crawl under the car to do my annual oil change and lube, another club member came to my house and did all the dirty work for me.

Here's another observation I've made. I always thought that sports car owners were outgoing, talkative, even overbearing, the reverse of what I am. When I joined the club, I noticed just the opposite – that many of our members were very quiet. Nice people, but quiet and I was the one that had to work to keep the conversation going. Some of us are talkers. Some of us aren't. Some of us find it hard to start talking with a total stranger. But, at a club get-together or especially a car show when someone comes up to admire our shiny car and asks a question, it's our opportunity to open up and gab with a total stranger with a common interest. Whether we realize it or not, most of us are walking encyclopedias about our car, it's history, what we love most about it (and those annoying little things that we live with and will get around to fixing one of these days), even the history of the company that made it. We also may discuss the evolution of human beings (i.e., "Those Brits back in the 50s and 60s must have been a lot smaller than people are today.") Our cars get us up, out, moving, and talking to people.

Here are my thoughts on car shows. I do three or four shows a year, usually baking in the hot sun all day. I arrive early and detail my car so that it'll look spotless for the judges and the hundreds of people who'll pass by and take a look. It's nice to take home a trophy or a plaque. It tells us that the judges liked our car best, or maybe second best. It's disappointing to not win, but then I remind myself that a physical award is not the ultimate prize. It's about getting together with people, talking, having fun, learning from others, and discovering common interests. I suppose we're not all that different than people who go to quilt shows, flower shows, woodworking shows, even koi fish shows.

And here's a suggestion if that award really is important to you. Don't be afraid of the people judging your car. They're car people too. They're most likely not an expert on your car, so tell them about it just like you would anyone else who comes up to look. Chances are they're just as curious. Open the hood and trunk so they know you're not hiding anything. Smile, so *you* come off as non-threatening. By chatting-up with them, you're establishing a bit of an inter-personal bond, a momentary friendship, that may gain you a few points.

So, what is a club about? It's not just about having fun at meetings, doing car shows and pub runs, and celebrating good times together. It's also about sharing sad times, and drinking a toast to friends who are no longer with us. It was the long, scenic funeral procession of MGs to the cemetery, hoping Joe Hooker enjoyed his final ride, even if it wasn't in his TD, and then Bob Farrell a few years later. It's driving together to an MG gathering half way, or even all the way, across the country. It's helping a club member replace a head gasket on the side of the highway in the middle of nowhere (which included finding an auto parts store in a small town that had gasket material and a compression tester in stock).

A club is all about making friends and reaching out to people. The cars are just the excuse that we use.

What is Authentic? What is Original? What is Over-Restoration?

One of the questions I'm often asked at car shows when showing my 1950 MG TD is "Is it original?" My answer always is "No, but it is authentic." I then explain that once the car has its first oil change, it no longer is original. I then go on to explain how 'authentic' is different from 'original.'

Over the years I've seen plenty of British and other cars at car shows. The way I look at it, I see three categories regarding the extent of restoration: unrestored, authentic restoration, and over-restored. You may have more categories. I'll expand on each. I won't even discuss the few MG TDs made into hot rods or drag racers, as I have to hold down my lunch when I think about them.

I've seen many unrestored cars at shows and cruise-ins. None were shabby, but usually showing their age, even when cared for well, showing their patina of age, which one would expect for a vintage car. Even when garaged and driven lightly, age still takes its toll. That doesn't mean they're ready to be parted out, just in need of some serious work. Paint and interior may be worn, engine may not be putting out anything near full power and even may be putting out a bit of burned oil with the exhaust, and the transmission may be shifting roughly. The things that stop working all together may be the only items that get any real attention. At a show several years ago, I parked next to a really rough-looking MG TD. Chatting with the owner amongst the row of sparkling TDs, he smiled and said "Watch this...I can do something none of you can do with your cars..." He proceeded to shake his soft drink can and then place it on the hood (Excuse me...bonnet, for you British purists). As the contents foamed up and spilled out of the can, he smiled again and said "Doesn't bother me a bit." I'm sure he loved his car as much as I love mine. Maybe he even has more fun, because he doesn't worry about every little scratch or blemish.

My car is the second category: authentic restoration. An authentic restoration should make the car look like it did when it came off the assembly line, but it may include period-appropriate accessories. My car's previous owner, who restored it, took great pains to keep as many original parts as could be saved. He had to replace the wood frame and the interior, as the originals were too far gone, but all the metal is original, the frame and engine numbers match, and neither the engine nor the transmission had ever been rebuilt. Other worn parts were replaced with modern reproductions of the originals. He even welded over where another previous owner had drilled through the fenders to mount side-view mirrors, then polished the metal leaving no trace of the repair. (The only give-away is the blob of metal on the under side of the fender where it isn't visible.) The color isn't original, either. It is red, but I think a deeper red than was an original MG color, based on verbal descriptions. The original color can't be reproduced exactly because the dyes used years ago aren't available today. Colors fade over time and with exposure to sunlight (red is the worst), and color samples from period sales brochures may not even be that accurate. Unless someone has original paint samples that have been stored in a cool, dark location, I doubt many original colors can be reproduced exactly through the use of laser-scanning analysis. Of all the red TDs I've seen, only one matches my color, as the owner saw my car (years before I owned it) and had his painted the same color (or maybe it was the other way around...I don't remember). MG TDs came in five or six different colors. They even could be ordered in primer only, so the owner could have it custom-painted. Thus, almost any color, even two-tone, could be considered authentic or period appropriate. Many owners used modern enamel paint instead of lacquer, and upgraded electronics to make them more reliable. Generally, these are upgrades that would not be noticeable. Many of us have replaced bias ply tires with radials. It throws the look off slightly, but only an antique British car enthusiast would notice. I added seat belts, but the color matches the interior and they have chrome buckles, and look period-appropriate. And one last thought... Sometime in its earlier life, my TD threw a rod through the oil pan. The owner welded a patch over the hole. It's still there, and I wouldn't think of replacing the oil pan with an undamaged one. It's part of the car's history. Of course, you have to crawl under the car to see it, and the rest of the car looks great!

My last category is over restoration. This is when a restorer makes the car better (in their opinion) then it looked when new. I've seen stereo systems and plush interiors to name a few. I've also seen paint jobs so shiny that I needed to put on my sun glasses when looking them over. Old cars, even when new, never sparkled like that. My MG TD came from the factory painted with lacquer. I've tried and tried, but I don't think one can put enough wax on lacquer or buff it enough to make it sparkle like a modern enamel/clear coat paint job. And, to be honest, to me it just doesn't look right, but when judging is by popular vote or the Mayor's Choice award, the sparkly car always will get their vote. There also is one car that shows up often that has everything in the engine compartment chromed. Even the bolts; even the hoses, which appear to be some sort of flex-metal, not rubber. The only things not chromed are the rubber belts, and I bet we'll see those replaced with something made with a chrome finish some day.

…and don't get me started about trailer queens that only get driven from the parking lot onto the show field, but if that is what gives pleasure to the owner... I've admired them, but think the owner is only getting half the pleasure that is to be had. In all it still comes down to the owner's objective. Why do you have it? To drive? To preserve the marque? To do both? I think the most important thing is the pleasure the owner derives from the car; to drive and enjoy these great cars as they were meant to be, whether they are rough or pristine.

Be a Safer Driver

Following the untimely and tragic deaths in 2021of our friend and MG Car Club member Bob Farrell, and then learning about the death of someone on their way home from British Invasion in Stowe, Vermont, I got to thinking about what I do to drive more safely, especially in my MG. I don't know any details of Bob's accident, but knowing Bob, I'm sure he was driving safely and doing everything properly. Classic cars lack most of the safety features we probably take for granted in our modern cars. In the early days of automobiles, it was assumed that since the vehicle was an inanimate object, any accident had to be the fault of the driver, not the vehicle design or the surroundings, and thus safety consisted of teaching safe driving habits. You've probably thought about the same things I'll discuss below, but here is my take on them.

Seat belts – If your car doesn't have them, add them, even if not equipped originally. Period appropriate belts are available and don't detract from the classic appearance you probably want to preserve. I doubt any judge at a car show would mark you down for having them, if installed nicely. I added them soon after purchasing my MG TD. They're airline-style belts with chrome buckles. At shows, rather than hide them, I cross them neatly across the seats. I secured them firmly to structure, and after owning my TD for twenty years, this modification is the only time I've drilled into original metal. I purposely added only the lap belt and not a shoulder strap, because being a convertible, the top end of the belt cannot be attached at a point above the occupant's shoulder. In a crash, a shoulder strap attached lower down can cause s serious spinal compression injury as your upper torso pivots forward. If you recall, convertibles pretty much disappeared in the late seventies or early eighties because of the lack of roll-over protection. When they reappeared about a decade later, the top of the door and door pillar were raised to be nearer the occupant's neck, allowing the shoulder strap to be mounted higher up, and the windshield frame was beefed up to serve as a roll bar. And I almost forgot to mention…if you already have them, use them.

Watch your speed – Modern cars are made to absorb energy in a crash; classic cars were not. Classic cars, especially convertibles, also have little to no roll-over protection. (Take note of where the top of your door is relative to your shoulder, and that your wind screen will provide little roll protection.) Will your car continue to travel straight ahead if you have to jam on the brakes or make a sudden yank on the steering wheel? Have you ever tried it to see what happens? I haven't and don't plan to try it.

Keep clear of other cars – Keep your distance from other cars, as best you can. While you can't control how close someone pulls up behind you for a close look, you can keep extra distance from the car ahead of you. This will give you a bit more reaction time and greater distance in which to stop if you have to hit the brakes. I especially hate to stop at a traffic light that is at the top of a hill. The TD will always roll back a foot or so as I gently engage the clutch, and I'm always afraid that the car behind me is way too close. I have two intersections near my home where I've found myself in this situation, and thus avoid them.

Make sure your brakes are working properly – If you're used to keeping extra distance and braking gently, you may have fewer than four operating brakes and not know it. A panic stop could result in your car getting into a skid and going out of control. I'm probably as gentle on my TD as any driver ever has been, and usually brake gently. A few years ago, I was driving home on a wet road (yes…I have been caught in the rain a few times) and approached a traffic light at the bottom of a steep hill. I thought I'd have enough time to make it through the intersection, but didn't. I hit the brakes and the car swerved about 45 degrees (well, it seemed like that to me) accompanied by an awful screech. I took the car to our local British car shop to have the brakes checked. It turns out I was driving with only one operating brake shoe. Obviously, since I always brake gently, I never realized that three brake shoes weren't working. That's probably the only big scare I've ever had in the TD. Lesson learned.

Tires – Make sure your tires are properly inflated – Our MGs are so light that the tires may still look like they're full of air, even with only ten pounds or so of pressure. Your car may drive O.K. while going straight, but probably will feel "squishy" on a corner or after a sudden yank on the steering wheel. And while I'm on the subject, consider switching to radial tires if you're still using old-style bias ply tires. It may change the "authentic" look just a bit, but I think the safety increase is well worth it. I switched about ten years ago and couldn't believe the difference. I bought a low-priced set, given that I only drive the TD about 400 – 500 miles per year and don't drive it aggressively. I'll probably never wear out the tires before they time out. Pulling out of the tire shop for the first time, it felt like I was driving a different car, one about twice as wide and twice as heavy. The tires really gripped the road, especially on a turn. This probably is because the radial layout of the steel belts cuts down on sideways flexing and a softer blend of rubber really grips the pavement. With my old tires, hitting a manhole cover in a turn resulted in the car skidding sideways a bit, probably not much, but enough to get my attention. No such sensation now. Being told that tires start to crack and should be replaced after seven years, I finally had them inspected after about ten. The technician said they looked great, probably because the car spends about 99.9% of its time in the garage, and the sun's ultraviolet rays are what cause most of the damage. He recommended having the tires inspected every year. Good advice.

Maintain situational awareness – This means keeping your eyes moving, not focused just on the road ahead, but all around you. Might someone pulling into an intersection or riding on a bicycle not see you? Could a child playing in their yard or a pet suddenly dart into the street? Anticipate what could go wrong and be ready to respond.

Check your exhaust system – Most of us don't think much about our exhaust system until we suddenly hear an unexpected and unearthly roar. Most of our classic cars don't seal that well, and exhaust fumes, even from a small leak or near the rear of the car, can enter the passenger compartment. For summer driving with the top down, this shouldn't be a problem, but for those chilly fall or spring drives with the top up, be cautious of the smell of exhaust…it might not be coming from the car ahead or behind you. When you're working under your car, be sure to do a visual check for any dark, sooty spots on the exhaust manifold, pipes, and muffler. There are numerous products to seal small holes that can give you maybe another year of service, or at least give you a bit of time to replace the defective part.

Keep your nose on alert – Your nose is pretty sensitive to smells, but the sensitivity fades as the smell lingers. Pay prompt attention to them. A sweet smell might be a coolant leak. Rubber burning? Might be an electrical short or a tire problem. Exhaust smell? See above paragraph. Ever driven with the parking brake on (no comment)? A burning smell might be the first thing that gives it away.

These are just a few ideas, the things I think about when I drive, but especially think about when I drive my MG. Please, plan ahead, be aware and be safe.

I Fixed It Myself

I don't think that many of our club members are professional auto technicians or restorers, but we all love to tinker (ya, right) with our cars. Isn't that half the fun of owning a classic car (well, maybe a quarter)? We've all fixed things, and we're all eager to pass along to others the knowledge we've gained. We may not know how to overhaul a transmission, but we've all found something amiss, identified the problem, and fixed it.

I've always bragged about how my 50 TD does not leak oil. Several years ago I arranged to take a friend for a ride. I pulled up on his brand new driveway, picked him up, and off we went for a spin. When we came back, I parked on the drive and we stood around and chatted for a few minutes. When I went to get back in, I noticed a big black puddle on his new, spotless white concrete. Then we saw the drops all along his driveway and down the street as far as we could see. Embarrassed as could be, I lifted the hood to check the engine, apologizing profusely, explaining that "**My MG doesn't leak oil**." We looked around the engine compartment and couldn't find oil anywhere. Worse, the dipstick didn't even show any oil, so it must have been dripping for a while!

I started the engine to see if I had oil pressure. This time we saw oil pouring out onto the drive. I lifted the hood again and quickly found the source…the banjo fitting where the oil returns from the filter to the engine block. Oil was shooting straight down…not dripping…shooting!

I was afraid to drive the car home at this point, not wanting to have to stop and add oil every mile or so. My friend was gracious enough to let me park the TD in his garage. He drove me home and we returned with my toolbox.

I removed the rigid steel pipe that leads from the oil filter to the engine. I then took the wire brush and bottle of dishwashing soap that I brought with me and tried my best to scrub away the oil spots from the driveway (what was left of them eventually faded away after a few months).

The cracked part looks is a U-shaped pipe with a brass banjo fitting brazed to each end. I took the part home and jury-rigged a test jig with which I could pressurize the pipe. Sure enough, I found air bubbles coming through the weld on the engine side. I took the pipe to a local shop, where they brazed the part for me while I waited. I also picked up four new copper gaskets.

When I reinstalled the pipe, I realized why the weld had broken. The oil filter assembly is mounted to the side of the engine with a strap. Oil travels from the oil pump to the filter, then back into the engine through rigid lines. The strap holding the assembly to the engine had come loose. Obviously, not only the weight of the assembly, but also all of the vibration loads and the shock loads from every bump in the road was going through the welds. We'll probably never know why the MG designers didn't use more flexible lines to make the design more tolerant to a loose oil filter clamp. (If they had thought it out better, it wouldn't be British...would it!)

Anyway, I got it all back together a few days later. I now have an additional item to check in my winter maintenance...make sure the strap holding the oil filter assembly is snug.

Another Crisis

As antique or classic car owners we learn to do a lot of the maintenance ourselves. It seems like there is always some little thing that needs fixing, or just a bit of tweaking or adjusting. Sometimes, just taking an item apart, inspecting (and finding nothing obviously wrong), and reassembling fixes the problem. I'm sure we've all been there!

It was the third winter that I owned my red 50 TD. During this season, I started it every month or so, long enough to let the engine warm up, and if the driveway was dry, I'd drive it in and out of the garage a few times. When I put it to bed that fall, it was running fine (relatively, of course, for a 52-year-old British car). I last drove it during late November. A month later, I couldn't get it to start. It turned over, but not even a sputter. Concerned, I let it sit for an hour, then came back and tried again. This time, it started, but ran terribly. I assumed it just needed to get out on the road to burn off whatever crud had accumulated somewhere in the engine.

The same thing happened when I ran it in late January. It wasn't until early March that I finally got the car out on the road on a sunny, dry day. At least, I tried to get it out on the road. As I drove through my plat, I realized I was getting almost no response from the gas pedal. I got up to 25 mph and into third gear, but couldn't hold speed on even a slight incline. I drove about five miles or so, long enough to get the engine up to temperature, but nothing changed. If I pushed in the mixture knob, which also lowers the idle RPM, the engine would stall. My best guess was that I was running on one cylinder.

While I usually can sort out a mechanical problem if I can see the moving parts, I find electrical problems more difficult. I pulled the spark plug wires one at a time to see if the problem could be fouled plugs. After checking wires 1, 2, and 3, there was virtually no change in how the engine idled. I was getting a good spark jumping from the wire to the engine, so I concluded each plug was getting electricity. But when I pulled number 4, the engine quit. This at least confirmed my suspicion that it was running on only one cylinder. Guessing that perhaps the plugs were fouled, I pulled and cleaned them one at a time to eliminate the chance of reconnecting the wires in the wrong order. This helped it start a little easier, but the engine ran no differently. I even replaced the plugs with new ones, and cleaned the points and contacts in the distributor cap, but still no improvement. At this point, I called my friend Joe Hooker.

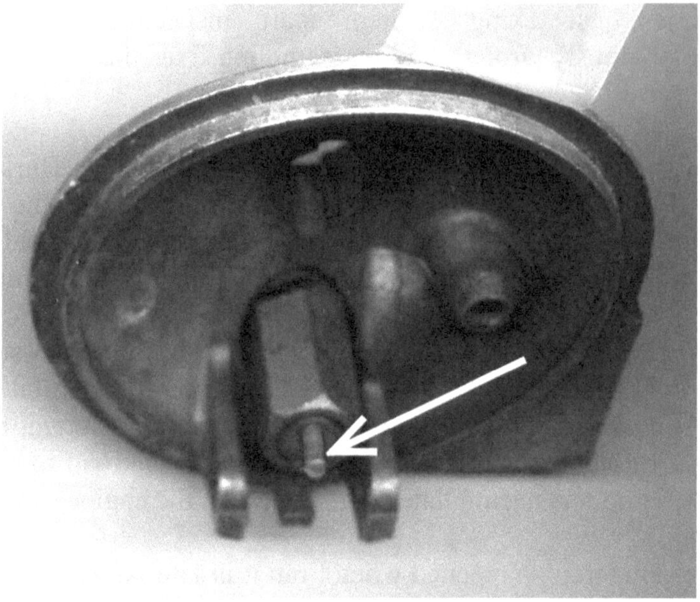

Inside of the carburetor bowl cover. Note arrow pointing to the pin that stuck. Pin is in fully extended position, which will allow fuel to flow.

Joe came over and quickly confirmed that something was wrong. He re-checked everything that I already had checked, but nothing changed. He then started looking at the fuel system. We confirmed that the pump was running fine, even pulled the fuel line from the rear bowl to make sure it really was pumping gas (it was). We put it back together, but still no improvement. I accidentally left the ignition switch on, and as we chatted for a few minutes, we noted that the fuel pump ticked about every thirty seconds. Normally, my pump hardly ever ticks once the fuel lines are pressurized (i.e., once the bowls are full, which happens fairly quickly, the pump no longer pumps). As we talked, neither of us thought much about the ticking.

Joe and I took the car out one more time. This time, it accelerated normally. But as I got up through third gear the power disappeared again. We speculated that one or both of the bowls was not filling properly. Apparently, in the time that we were talking and the pump kept ticking, the bowls had filled slowly, providing enough gas for the car to accelerate normally for a few hundred yards, but then one of the bowls went dry.

After Joe left, I repeated the pump test on the front bowl as Joe had done on the rear one. There was plenty of gas flowing and the filter was clean, so I reconnected the fuel line and pulled the cover from the bowl. I found the bowl to be almost empty, meaning the valve obviously wasn't opening to let the fuel fill the bowl! I turned on the ignition, but only a drop or two of gas flowed from the valve every few seconds, confirming my suspicion. (I guess I should explain…there's a float in the bowl that pushes on the valve. When fuel level drops, the float drops and the valve opens to let gas flow to refill the bowl.) At this point, I jiggled the pin on the bottom of the valve (see photo), and gas gushed out! I quickly shut off the ignition and jiggled the pin several times. It now moved freely, so I put everything back together. There must have been a speck of dirt making the pin stick. The valve has worked properly ever since.

Joe…thanks again for the help. We all miss you. This is what a car club is all about!

Automotive and Mechanical Stuff

Horsepower – What Does it Mean?

Editor's note – this article is based on one by Marshall Brain that I found on the internet. I added to it to make it my own, but give him credit for much of the text, as he'd already explained it very clearly.

Chances are you've heard about horsepower. Just about every car ad mentions it. But what exactly is horsepower?

But first, we need to understand a few other terms, namely work, as a scientific measurement, not just something you do and maybe get paid for, as in working a job. If you push on an object, you are exerting a force. If the object moved, the force was applied over a distance, and this is work in a scientific sense. Work is the product of the force times the distance over which the force is applied, and is measured in terms of foot-pounds. Power is the amount of work performed in a given amount of time, commonly measured as foot-pounds per minute.

The term horsepower was invented by the engineer James Watt, who lived from 1736 to 1819, and is most famous for his work on improving the performance of steam engines. The story goes that Watt was working with ponies lifting coal at a coal mine, and he wanted a way to talk about the power available from one of these animals. He found that, on average, a pony could lift 220 pounds a distance of 100 feet in one minute, i.e., they did 22,000 foot-pounds per minute. He then arbitrarily increased that number by 50 percent, figuring that's how much a full-sized horse could do, and pegged the measurement of horsepower at 33,000 foot-pounds of work in one minute. It is that arbitrary unit of measure that has remained through the centuries and now appears on your car, your lawn mower, your chain saw and even in some cases your vacuum cleaner.

As stated, in Watt's judgment, one horse can do 33,000 foot-pounds of work every minute. So, imagine a horse raising coal out of a coal mine as shown above, raising 330 pounds of coal 100 feet in a minute. But can the horse lift 33 pounds of coal 1,000 feet in one minute, or 1,000 pounds 33 feet in one minute? Possibly, with some arrangement of pulleys, and you can make up whatever combination of feet and pounds you like, but as long as the product is 33,000 foot-pounds in one minute, you have one horsepower.

You can think of many combinations that surely aren't practical or even possible, but still would be true to the definition. You probably can imagine that you would not want to load 33,000 pounds of coal in the bucket and expect the horse to move it 1 foot in a minute because the horse couldn't budge that big a load. You probably also can imagine that you would not want to put 1 pound of coal in the bucket and ask the horse to run 33,000 feet in one minute (which translates into 375 miles per hour). Again, one horsepower, but most probably impossible to achieve.

You probably hear the term torque frequently associated with horsepower. Torque is a measurement of the twisting capability of the engine. It's related to horsepower, but different engines may produce different torque at the same power and RPM, but I won't get into it at this time.

There are other ways to express power besides horsepower. For example:

- 1 horsepower is equivalent to 746 watts (an awfully bright light bulb, but not much for a hair dryer). So if you took a 1-horsepower horse and put it on a treadmill, it could operate a generator producing a continuous 746 watts. This may seem like an oddball number, but it's based on metric units, and gives us this number when converted to the English system.

- 1 horsepower is equivalent to 2,545 BTU (British Thermal Units). If you took that 746 watts and ran it through an electric heater for an hour, it would produce 2,545 BTUs, where one BTU is the amount of energy needed to raise the temperature of 1 pound of water 1 degree F…again, if you use kilogram and degrees C, it comes out to a nice round number).

- Air conditioners are rated in tons, which is how many tons of water the unit can freeze in one hour starting at some given temperature. This measurement also can be converted to horsepower, but I won't bother introducing any more confusing conversion factors.

So, how do engineers calculate the anticipated power from a new engine design? I don't know! I couldn't find anything on it. As best I can figure, they make a good guess based on past experience, but don't know for sure until they actually test it. The story has it that the Wright Brothers, after determining the horsepower they'd need to get their airplane to fly, tested their engine. They took a bucket of sand of a known weight, attached a rope and connected it to the engine's shaft. Then they measured how long it took the engine to raise the bucket a given height. You can't get much simpler than that.

Let me leave you with a little story. Back when I was in high school, we had an essay question on a physics exam that went something like: "A physics teacher hired a few neighborhood kids to move a large boulder in his yard. The kids pushed and pushed for hours, sweating gallons of sweat in the hot sun, but the rock never budged. The physics teacher then refused to pay the kids because he claimed that they did no work. Was he correct in his refusal?" Being a public school, not a private one at which they can teach morals, my answer certainly concluded that from a physics standpoint, they did no work, and thus didn't earn the pay they'd been promised. I guess that if there's a moral to the story, it is don't ever work for a physics teacher.

Understanding Engines

Horsepower, RPM, Octane, Compression Ratio… Here's what it all means.

My 1963 REO snow thrower died a few winters ago. Actually, it died years ago, but I kept it alive, probably spending more time reviving it each winter than I spent actually plowing snow. I gave in and bought a big Troy-Built, which I figured would be overkill for our Dayton winters, but our occasional winter blizzard still gives me something to look forward to. While researching which model to buy, I noticed that they often didn't give the engine's horsepower, just the displacement. I figured there should be a simple way to convert. But, while in general more engine displacement means more power, I learned that the relationship is much more complex than just the displacement. Thanks to the miracle of the internet, here's what I found.

First of all, what exactly is "horsepower?" Horsepower is a measure of work performed per unit of time. Work is a force exerted over some distance. For example, if you push a rock with 50 pounds of force and push it for 10 feet, you did 500 foot-pounds of work. If you did this in ten seconds, then you did 50 foot-pounds of work per second. Sometime in the past, it was determined that a horse can do about 550 foot-pounds of work per second, and called it a horsepower. Thus in pushing the rock, you generated just under 1/10 horsepower. (BTW, if the rock hadn't moved, then you did no work, regardless how much you sweated, because the force wasn't exerted over a *distance*!)

I'll skip the basic theory of how a four-cycle gasoline engine operates. I assume we all understand the basics. We all realize that fuel contains energy, and the more fuel and air we can get into the engine to burn, the more power the engine can produce. There is an obvious relationship between displacement and power, but there are many other factors to consider. Before looking at the engine's displacement, first let's consider the intake stroke. As the piston moves down the cylinder during the intake stroke, it creates a partial vacuum, sucking air and vaporized fuel through the intake valve.

The faster the piston moves, the faster the air has to flow through the intake valve.

Engineers have observed that air cannot really flow through the intake at speeds greater than about 650 feet per second. This appears to be a critical speed: trying to pull it through at a higher speed takes more power than you get from the resulting combustion. (But, depending on the design of the intake, that speed could be as high as 710 feet per second, or as low as 600 feet per second.) In general, more RPM means more power, because more fuel/air is being drawn into the engine, until the intake speed hits 650 feet per second. At that point the engine reaches its maximum speed because it just can't suck fuel and air into the cylinder any faster. Thus the size of the intake valve limits the maximum RPM and thus the maximum power. Increasing the size of the intake valve allows the engine to rev faster, because more fuel and air can enter the cylinder before the mixture reaches the 650 foot per second limit.

Other things being equal, the engine with the biggest intake valves produces the most power, not necessarily the engine with the most cubic inches of displacement. For example, a 180 cubic inch Formula 1 engine with 32 square inches of intake valve area generates 720 horsepower, but a 427 cubic inch Ford racing engine with only 28 square inches of valve area generates only 600 horsepower, despite having nearly 2½ times the displacement! But, to generate that maximum power, the Formula 1 engine spins at 13,500 RPM while the Ford engine only 6800 rpm. While the Formula 1 engine probably is lighter, because of its lower displacement, it needs a transmission capable of gearing down the RPMs, and, an engine turning that fast will have different lubrication and maintenance issues.

The size of the exhaust valve usually is not a limiting issue. During the intake stroke, fuel/air is sucked in from the atmosphere. After combustion, the exhaust now is at several hundred atmospheres of pressure and will shoot quickly out of any size opening.

What about octane? Remember those WWII movies with the tanker carrying a load of high-octane aviation fuel? It always sounded more dangerous than if they were only carrying regular fuel, right? High performance engines tend to use high octane gas, but does using high octane gasoline increase power? No. and here's why. All gasoline, regardless of its octane rating, has the same amount of energy per unit of volume. So, what does octane mean?

Octane is a measure of the resistance of the fuel to start burning by itself (autoignite) before the spark plug fires. It might seem odd that slower-burning fuels are used in more powerful engines (we'll get to this in a minute), since high octane fuels are harder to ignite than low octane fuels. Recall that the fuel must ignite at a precise time. An ignition too early, while still in the compression stroke, will try to push the piston back down the cylinder, causeing strain on the crankshaft and other parts. This strain is the source of the characteristic 'ping' noise. Autoignition reduces power output because much of the energy is absorbed as strain and heat in various parts of the engine rather than being converted to torque at the crankshaft. (BTW –autoignition, ping, and knock are actually slightly different phenomena, but the terms usually are used interchangeably)

At this point, let's divert for a monent to the engine compression ratio.

Compression ratio of an internal-combustion engine is the ratio between the volume of the cylinder (and thus the combustion chamber) when the piston is at the bottom of its stroke, and the volume of the combustion chamber when the piston is at the top of its stroke. Compression ratios of 10:1 are typical, with high-performance engines reaching 13:1.

A high compression ratio is desirable because it allows an engine to extract more mechanical energy from a given mass of air-fuel mixture. High ratios place the available oxygen and fuel molecules into a reduced space, causing better mixing and evaporation of the fuel droplets. Thus they allow increased power at the moment of ignition and the extraction of more useful work from that burst of power by expanding the hot gas to a greater degree. So here's the downside; the more the fuel/air is compressed, the hotter it gets, making it easier for it to ignite by itself before the spark plug fires. Thus the higher octane is needed to prevent the mixture from autoigniting.

Burning fuel with a lower octane rating than required by the engine often reduces power output and efficiency. If the engine begins to detonate (knock), it reduces power and efficiency for the reasons stated above. Many modern car engines feature a knock sensor – really nothing more than a small microphone that detects knock, and then sends a signal to the engine control unit to retard the ignition timing. Retarding the ignition timing reduces the tendency to detonate, but also reduces power output and fuel efficiency.

Higher compression ratios will, however, make gasoline engines subject to engine knocking, also known as detonation, if poor quality fuel is used. This can reduce efficiency or damage the engine if knock sensors are not present to retard the timing. However, knock sensors have been a requirement of the OBD-II specification used in 1996 model year vehicles and newer.

So, buying cheaper, lower-octane gas than your engine needs will cause it to lose power and could damage the engine, not really saving you money in the long-run. But, buying more expensive, higher octane than your car needs won't make it run any better!

Most fuel stations have two storage tanks, even if they offer 3 or 4 octane levels. They just mix the higher and lower octane fuel. Purchasing premium simply means more fuel from the higher octane tank. The detergents in the fuel are the same.

Timing referrs to the exact time that the spark will occur relative to piston position and crankshaft angular velocity. Setting the correct ignition timing is crucial for proper engine operation. By igniting the air-fuel mixture at the correct time, maximum pressure in the cylinder will occur, allowing the ignited mixture to push the piston down the cylinder with the greatest force. Timing can be either "on top dead center," "advanced," or "retarded." Ideally, the time at which the mixture should be fully burned is about 20 degrees after top dead center. Spark advance or retardation is required because of the different amount of time available to burn the air-fuel mixture at various RPMs. As RPM increases, there is less time for combustion, so igniting the mixture a bit earlier will allow the mixture to burn more fully. But, if the ignition spark is advanced too much, the rapidly expanding air-fuel mixture actually can push against the piston still moving up, causing detonation and lost power. If the spark occurs too retarded relative to the piston position, maximum cylinder pressure will occur after the piston is already traveling too far down the cylinder. This results in lost power, unburned fuel, and high emissions.

In older, non-computer-controlled gasoline engines, timing is controlled inside the distributor as it distributes the ignition spark to each cylinder. By using rotating weights and springs, centrifugal force causes the contacts inside the distributor to move forward (this is called centrifugal advance), so that the spark occurs sooner. Many of these engines also use a vacuum advance that adjusts the spark during "deceleration" and/or low load "coast" independent of the centrifugal advance. (editor's note – there is no such thing as "centrifugal force," but the term has become established in common use. Don't get me started on this.)

Dwell controls how much electrical energy is delivered to each spark plug. At ordinary engine operating speeds, the points (located in the distributor) open and close a couple of hundred times per second, the exact number depending on the number of cylinders and the engine RPM. The points need to be closed for an appreciable time in order to build up the maximum electrical charge in the ignition coil.

The period of points closure is specified by the ignition system designer and is typically expressed as degrees of distributor rotation. In a four cylinder engine, the angle between each ignition contact is 90° and the period of points closure or "DWELL" is usually a bit over 45° of distributor rotation. In a six cylinder engine, the contacts are 60° apart and the dwell time is 30° to 35°.

The dwell is adjusted by setting the points gap to a specified distance. A narrower gap gives more dwell and a wider gap gives less. Taking it to extremes, excessive dwell makes the points close too soon after opening, cutting off the coil before it delivers all its energy. Too little dwell gives the coil insufficient time to build up to the maximum.

Both conditions give a weak spark which gets even weaker as the engine RPM increases, producing misfiring at normal operating speeds. The dwell, as well as spark plug gap, have an effect on ignition timing. The later the points open, the later the spark comes and has the same effect as retarding the timing. The earlier the points open, the sooner the spark comes and advances the timing. That is why timing is set after the dwell when performing a tune-up. And, don't forget to check the gap on your spark plugs before you install them!

Air-fuel ratio (AFR) is the ratio of the mass of air to fuel. When there is exactly the correct amount of air needed to burn all the fuel, the mixtue is called a stoichiometric mixture. For gasoline, the stoichiometric air/fuel mixture is approximately 14.7 times the mass of air to fuel.

In theory a stoichiometric mixture has just enough air to completely burn the available fuel. In practice this is never quite achieved, due primarily to the very short time available for combustion during the power stroke before the exhaust valve opens, and also the cooler temperature at the cylinder wall which slows the burning process near this surface. Most of the combustion process completes in approximately 4-5 milliseconds at an engine speed of 6000 rpm., during which the crankshaft moves through about eighty degrees of rotation.

Any mixture less than 14.7 to 1 is considered to be a rich mixture, any more than 14.7 to 1 is a lean mixture. Lean mixtures produce hotter combustion gases, so much so that pistons can melt as a result. Rich mixtures produce cooler combustion gases, but also more carbon monoxide, because less gas is converted to carbon dioxide. Lean mixtures produce less power. Similarly, rich mixtures return poorer fuel efficiency.

A stoichiometric mixture burns very hot and can damage engine components if the engine is placed under high load at this fuel air mixture. Due to the high temperatures at this mixture, detonation of the fuel air mix shortly after maximum cylinder pressure is possible under high load (referred to as knocking or pinging). Detonation can cause serious engine damage as the uncontrolled burning of the fuel air mix can create very high pressures in the cylinder. As a consequence, stoichiometric mixtures are only used under light load conditions. For acceleration and high load conditions, a richer mixture is used to produce cooler combustion products and thereby prevent detonation and overheating of the cylinder head.

That's about everything I dug up about engines. So, how much horsepower does my 179cc engine produce? I don't know, but it sure blew up a blizzard of snow, especially when going through sixteen-inch drifts, with barely a hint that it even was working hard. I guess it has enough power.

One last nit-to pick while you're reading this. All too often I hear the terms "engine" and "motor" used interchangeably. This is another stomach-churner, as is "centrifugal force." An engine is powered by an expanding gas, while a motor is powered by magnets repelling each other. Unless you drive an all-electric of a hybrid vehicle, the engine powers the wheels while a motor powers accessories such as the fan and windshield wiper.

Early Developments in the Internal Combustion Engine

I've always wondered which came first, the internal combustion engine or the gasoline that fuels it. Were they both invented together, or was gasoline developed because the engine required a specific type of fuel?

Many scientists and engineers contributed to the development of internal combustion engines, going at least to 1794. As an engineer, my best guess is that the idea was based on the workings of the steam engine, which used high-pressure steam to push a piston down the inside of a hollow cylinder and then turned this linear motion into rotary motion by means of a crankshaft. These early internal combustion engines ran on anything that could be pumped into the combustion chamber, which included liquids, gasses, and even dust (my guess is that this was coal dust). As early as 1807, an internal combustion engine powered a boat on the Saône river in France. The same year, a Swiss engineer built an internal combustion engine ignited by an electric spark. A British engineer demonstrated his engine in a land vehicle in Kent in 1826, possibly being the first "automobile."

These early engines were called atmospheric gas engines. Prior to the invention of the spark plug, as best I can figure, fuel and air were ignited in a continuously-burning flame outside the combustion chamber. The burning fuel was drawn into the combustion chamber through a slide valve on the downward stroke, at atmospheric pressure. At some point, the piston closed the opening and the combustion continued to push the piston down as a result of the gasses expanding from the combustion. As best I can figure, this appears to be a two stroke engine, since there was no compression stroke to compress the fuel prior to ignition

Research continued through the 1800s to develop an engine that actually was useable on a commercial level. Up to this point, most of the engines were large and heavy, and could not exceed 200 rpm due to their ignition and induction systems. British engineer Edward Butler constructed the first petrol (gasoline) internal combustion engine. Butler also invented the spark plug, ignition magneto, coil ignition and spray jet carburetor. By 1883, an engine built by Gottlieb Daimler achieved over 900 RPM. In 1885 Daimler constructed what is generally recognized as the prototype of the modern gas engine: small and fast, with a vertical cylinder, and used gasoline injected through a carburetor. In 1889 Daimler introduced a four-stroke engine with two cylinders arranged in a "V," having a much higher power-to-weight ratio. With the exception of electric starting, which would not be introduced until 1924 (I know...I gave a different year elsewhere, as the information available varies), most modern gasoline engines are descended from Daimler's engines. Other contributors whose names you might have heard of (if you ever took a course in thermodynamics, anyway) are George Brayton, Nikolaus Otto, Karl Benz, and Rudolph Diesel (obviously inventor of the Diesel engine, which used the heat from compressing the fuel/air mixture to ignite it).

So, getting back to my original quest, how did gasoline get hitched up to the internal combustion engine? Which came first? It turns out...it was gasoline. Gasoline was considered a useless by-product of the refining of crude oil to make kerosene, which was a standard fuel used for lamps through much of the 1800s. Oil refining of the day used a relatively simple distillation process that separated crude oil into portions, called fractions. Using a closed container, the oil was heated and the vapors collected as they condensed at different temperatures. Gasoline had a light weight and lower boiling point than kerosene, so it boiled off first and generally was discarded. (I assume this was because it burned too easily and thus was unsuitable for use in lamps.)

As the engines evolved, engineers realized that they ran best on lighter fuels such as gasoline (remember, the fuel has to ignite and burn in only a few thousandths of a second), but the refining process could produce only about 20 percent gasoline from a given amount of crude oil. A gasoline shortage soon developed, even as the search for oil pushed into Indiana, Oklahoma, and Texas. For this reason, most early automobiles were electric or steam driven. Of the 4,200 cars built in the United States in 1900, only one-fourth used internal combustion engines.

As the demand for gasoline soared, chemical engineers worked on methods to extract more gasoline from the crude oil. Two engineers at Standard Oil of Indiana, William Burton and Robert Humphreys, came up with the idea to distil the kerosene under pressure. This caused the heavier kerosene molecules to "crack" into lighter gasoline molecules. The thermal cracking process produced twice the amount of gasoline, or 40 percent, from crude oil. Further development added chemicals to the mix to act as catalysts, yielding higher amounts of gasoline, while the refining process worked at lower pressures (which, I'd assume, increased safety at the refinery, and possibly made the equipment lighter and less expensive).

So basically, that's how we got to where we are. I won't get into further developments of engines or gasoline, such as the addition of lead to make the engine run smoother, or other chemicals that replaced the lead.

Your Engine's Cooling System

Back about 1971, I was attempting to fix a problem on my 1967 Camaro's 327 V-8 engine, and snapped of the only remaining bolt that held the pulley wheel that powered all the engine-driven accessories. Being new in town, I called the only Chevrolet dealer and was told that it was fine to drive the fifteen miles to their shop—the engine would overheat, but as long as steam kept coming out, the engine was cooling by boiling the coolant. Good advice? Probably not, but I did it anyway and the broken bolts were repaired and the trip caused no apparent damage to the engine.

Jump ahead to 2009. I was driving my aluminum block V-6 Olds Silhoutette when the temp idiot lite illuminated. I pulled to the side of the freeway for a few minutes to let the engine cool, then drove about a mile until I could pull off the freeway at an exit and call for a tow truck. Major mistake. By the time I pulled off, the engine was running so roughly that I only could maintain about ten mph. In that one mile, that high temperature warped both heads and did other internal damage, resulting in $3500 in repairs.

It's the cooling system's job to keep the engine from overheating. Did it fail to do its job on my Olds? By 1971 standards, yes, but by 2009 standards, no. To produce their power efficiently, while keeping the weight down, modern engines are designed to run at very specific design points. Drift away from any one of these critical conditions, such as temperature, even for a few minutes, and major damage can result. So, here's what I know about your car's cooling system, and how each of its components contribute to this job.

Gasoline engines are not very efficient at turning chemical energy into mechanical power. Most of the energy in the gasoline (perhaps 70%) is converted into heat. There are three ways for this excess heat to be removed: by exiting through the tailpipe with the exhaust gases, by travelling through the metal block and radiating into the air, and by being carried off through the coolant flowing through the engine. The cooling system on a car driving down the freeway dissipates enough heat to heat two average-sized houses!

The primary job of the cooling system is to keep the engine from overheating by transferring this heat to the air, but the cooling system also has several other important jobs.

The engine in your car runs best at a fairly high temperature. When the engine is cold, components wear out faster and the gasoline burns less completely, emitting more pollution. So another important job of the cooling system is to allow the engine to heat up as quickly as possible, and then to keep the engine at a constant temperature.

The engine runs best when its coolant is about 200 degrees Fahrenheit (93 degrees Celsius). At this temperature, the combustion chamber is hot enough to vaporize the fuel completely, providing better combustion and reducing emissions. Also, the oil used to lubricate the engine has a lower viscosity (it is thinner), so the engine parts move more freely and the engine wastes less power moving its own components around. And, metal parts wear less.

A typical 4 cylinder vehicle cruising at around 50 miles per hour produces 4000 explosions per minute inside the engine. Temperatures in the combustion chamber can reach 4,500 F (2,500 C), so cooling the area around the cylinders is critical. Areas around the exhaust valves are especially crucial, and most of the space near the cylinder head is filled with coolant. If the engine goes without cooling for very long, it can seize. When this happens, the metal can get hot enough for the piston to weld itself to the cylinder.

The modern cooling system has not changed much since the 1920s. While it has become infinitely more reliable and efficient at doing its job, the basic cooling system still consists of liquid coolant being circulated through the engine, then to the radiator to be cooled by the air stream coming through the front of the vehicle.

There are two types of cooling systems found in automobiles: liquid cooled and air cooled. Air cooled engines are found on a few older cars, like the original Volkswagen Beetle, the Chevrolet Corvair and a few others. Many modern motorcycles still use air cooling, as do light aircraft and yard and garden equipment, but for the most part, automobiles and trucks use liquid cooled systems. I'll concentrate on these.

The cooling system uses a pump to circulate the coolant through passages inside the engine block and heads. A radiator cools the coolant, a thermostat controls the temperature of the coolant, a radiator cap controls the pressure in the system, and some plumbing consisting of interconnecting hoses transfers the coolant from the engine to the radiator and also to the car's heater system where hot coolant is used to warm the vehicle's interior.

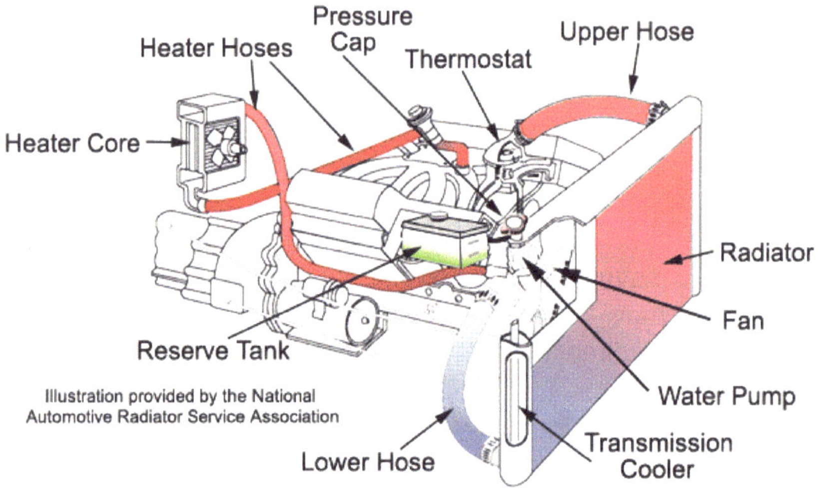

The cooling system pumps the liquid coolant through passages in the engine block and heads. As the coolant flows through these passages, it picks up heat from the engine. The heated fluid then makes its way through a rubber hose to the radiator on the front of the car. As it flows through the thin metal tubes in the radiator, the hot liquid is cooled by the air flowing past the metal tubes. Once the fluid is cooled, it returns to the engine to absorb more heat. The pump keeps the fluid moving through this system of plumbing, the radiator, and the internal engine passages.

A thermostat is located between the engine and the radiator to make sure that the coolant heats quickly to a pre-set temperature. When the coolant is below this temperature, the thermostat blocks the coolant flow to the radiator, forcing the fluid instead through a bypass directly back to the engine to absorb more heat. The coolant will continue to circulate like this until it reaches the design temperature, at which point the thermostat will open a valve and allow the coolant to flow through the radiator.

In order to prevent the coolant from being heated greater than the normal boiling point, the cooling system is pressurized. Under pressure, the boiling point of the coolant is raised considerably. However, too much pressure will cause hoses and other parts to burst, so a system is needed to relieve pressure if it exceeds a certain point. This is the radiator cap's function. The cap is designed to release pressure if it reaches the specified upper limit that the system was designed to handle. Prior to the '70s, the cap would release coolant to the pavement. Since then, a coolant recovery tank was added to capture any released fluid and store it temporarily in a reserve tank. This fluid then is sucked back into to the cooling system as the engine cools.

The capacity of the system, i.e., the size and location of the internal channels in the block, and the sizes of the radiator, hoses, and pump, is engineered for the type and size of the engine and the work load that it is expected to undergo. Obviously, the cooling system for a larger, powerful V8 engine in a heavy vehicle will need considerably more capacity then a compact car with a small 4 cylinder engine. On a large vehicle, the radiator is larger with many more tubes for the coolant to flow through, and also wider and taller to capture more air flow.

The coolant that circulates through the engine and associated plumbing must be able to withstand temperatures well below zero without freezing. It also must handle engine temperatures in excess of 250 degrees without boiling or breaking down. The coolant also must contain rust inhibiters and a lubricant.

The coolant in today's vehicles is a mixture of ethylene glycol and water. The optimum ratio for most geographic locations is fifty-fifty, i.e., one part antifreeze and one part water. In climates where the temperatures can go well below zero, it is permissible to have as much as 75% antifreeze and 25% water, but no more than that. Pure antifreeze is less efficient, will not work properly and can cause a boil over. (note – check the label carefully when buying antifreeze. Make sure it is the right type for your car, and also check if it already is diluted 50-50.)

Antifreeze is poisonous and should be kept away from people and animals, especially dogs and cats, which are attracted by the sweet taste. Ethylene glycol, if ingested, will form calcium oxalate crystals in the kidneys which can cause acute renal failure and death.

On most modern radiators, the tubes run either horizontally or vertically with a plastic tank on either end. On older vehicles, the core was made of copper and the tanks were brass. On newer cars, it is made of aluminum tubes and plastic tanks, and is much more efficient (apparently heat radiates from aluminum better than it does from copper), not to mention being cheaper to produce. Another component in the radiator for vehicles with an automatic transmission is a separate tank mounted inside one of the radiator's plastic tanks. Fittings connect this inner tank through steel tubes to the automatic transmission. Transmission fluid is piped through this tank to be cooled by the coolant flowing past it before returning to the transmission.

If the car has air conditioning, there is an additional radiator mounted in front of the normal radiator. This "radiator" is called the air conditioner condenser, which also needs to be cooled by the air flow entering the engine compartment.

Mounted on the back of the radiator on the side closest to the engine are one or two electric fans inside a housing that is designed to protect fingers and to direct the air flow. These fans are there to keep the air flow going through the radiator while the vehicle is going slow or is stopped with the engine running. In older cars without electric fans, every time the vehicle came to a stop, the engine temperature would begin rising. That is because the fan was connected to the front of the water pump and would spin at whatever speed the engine was turning. The speed was too low to maintain adequate airflow through the radiator. The driver might put the car in neutral and rev the engine to turn the fan faster while placing little load on the engine.

The electric fans are more efficient, since they force air past the radiator only when needed. They are controlled by the vehicle's computer. Note that when the air conditioning is turned on, the system will keep the fan running, even if the engine is not running hot. This is necessary for the air conditioning system to cool the car's interior.

While older cars, such as my MG TD, have unpressurized cooling systems, just about all systems on anything newer operates with the system under pressure. This allows the coolant to keep a higher temperature (and thus keep the engine at a more efficient operating temperature) without boiling. The radiator pressure cap maintains pressure in the cooling system up to a certain point. If the pressure builds up higher than the set pressure point, a spring loaded valve opens to release the pressure.

As coolant gets hot, it tries to expand, and the pressure in the cooling system increases. This is part of the design. When coolant is under pressure, the temperature where the liquid begins to boil is considerably higher. This pressure, coupled with the higher boiling point of ethylene glycol, allows the coolant to safely reach temperatures in excess of 250 degrees. When the cooling system pressure reaches the point where the cap needs to release this excess pressure, a small amount of coolant bleeds off. It could happen during stop and go traffic on an extremely hot day, or if the cooling system is malfunctioning. If it does release pressure under these conditions, there is a system in place to capture the released coolant and store it in a plastic tank that usually is not pressurized. Since there is now less coolant in the system, as the engine cools down, a partial vacuum is formed. The radiator cap on these closed systems has a secondary valve to allow the vacuum in the cooling system to draw the coolant back into the radiator from the reserve tank. There are usually markings on the side of the plastic tank marked Full-Cold, and Full Hot. When the engine is at normal operating temperature, the coolant in the reserve tank should be up to the Full-Hot line. After the engine has been sitting for several hours and is cold to the touch, the coolant should be at the Full-Cold line.

The water pump is a simple device that keeps the coolant moving when the engine is running. It is nothing more than an impeller mounted on a spinning shaft and usually driven by a belt on the outside of the pump body. The impeller draws the coolant in from the lower radiator hose and sends it under pressure into the engine block.

The thermostat is a valve that measures the temperature of the coolant and, if it is hot enough, opens to allow the coolant to flow to the radiator. If the coolant is not hot enough, the flow to the radiator is blocked and fluid is directed back to the engine. The bypass system allows the coolant to keep moving through the engine to balance the temperature and (hopefully) prevent hot spots. Because flow to the radiator is blocked, the engine will reach operating temperature sooner and, on a cold day, will allow the heater to begin supplying hot air to the interior sooner.

The secret of the thermostat lies in the small cylinder located on the engine-side of the device. This cylinder is filled with a wax that begins to melt as the coolant warms. A rod connected to the valve presses into this wax. When the wax melts, it expands significantly and pushes the rod out of the cylinder, opening the valve.

180 degree thermostats used to be the norm, but since the 1970s, thermostats keep the temperature of the coolant above 192 to 195 degrees. It was found that at these hotter temperatures, emissions are reduced, moisture condensation inside the engine is quickly burned off, and combustion is more complete, which improves fuel economy.

There is a mistaken belief by some people that if they remove the thermostat, they will be able to solve hard to find overheating problems. Not true!! Most often, the engine will never reach its desired operating temperature, or reach it very slowly. In some cases, by allowing uncontrolled circulation of the coolant throughout the system, it is possible for the coolant to move so fast that it will not cool properly as it races through the radiator, so the engine can run even hotter than before. Other times, on computer controlled vehicles, the computer monitors engine temperatures and regulates fuel usage based on that temperature. If the engine never reaches operating temperatures, fuel economy and performance may suffer considerably.

Freeze plugs are another important part. When an engine block is manufactured, a special sand first is molded to the shape of the internal coolant passages. This sand sculpture is positioned inside a mold and molten iron or aluminum is poured to form the engine block. When the casting is cool, the sand is loosened and removed through holes in the engine block casting leaving the passages for the coolant to flow through.

These holes need to be plugged, which is the job of the freeze plug. These plugs are press-fit in the holes and normally last the life of the engine with no problems. But there is a reason they are called freeze plugs. In the early days, many people used plain water in their engines and never replaced the antifreeze. When the water froze, the pressure of the expanding water forced the freeze plugs to pop out, relieving the pressure and often saving the engine block from cracking. Another reason for these plugs to fail was because they were made of steel and would easily rust through if the vehicle owner was careless about maintaining the cooling system.

Antifreeze has rust inhibitors in the formula to prevent this from happening, but those chemicals lose their effect after about 3 years, which is why antifreeze needs to be changed periodically. The fact that some people left plain water in their engines greatly accelerated the rusting of these freeze plugs.

Replacing the freeze plugs ranges from fairly easy to extremely difficult depending on the location of the affected plug. Freeze plugs are located on the sides of the engine, usually 3 or 4 per side. There also may be some on the back of the engine or on the heads. As long as you maintain the cooling system, you never should have to worry about these plugs failing on modern vehicles.

There are several rubber hoses that connect the cooling system's components. These hoses are designed to withstand the pressure and temperature of the cooling system. They are subject to wear and tear and eventually will require replacing as part of routine maintenance. Replace them if the rubber is beginning to look dry and cracked, or becomes soft and spongy, or if there is some ballooning at the ends. The main radiator hoses are usually molded to rout around obstacles without kinking. When purchasing replacements, make sure that they are designed to fit the vehicle. Flexible, generic hoses are available, and you'll have to determine their suitability by trial and error.

Proper maintenance of the cooling system is very important to the life of the engine. An engine that is overheating will quickly self destruct, or at least do major damage (remember my Olds Silhouette).

The most important maintenance item is to flush and refill the coolant periodically. This is because anti-freeze has a number of additives that are designed to prevent corrosion. This corrosion tends to accelerate when several different types of metal interact with each other. The corrosion causes scale that eventually builds up and begins to clog the thin metal tubes in the radiator and heater cores, eventually causing the engine to overheat. The anti-corrosion chemicals in the antifreeze prevent this, but they have a limited life span.

Most antifreeze used in vehicles is green in color and should be replaced every two years or 30,000 miles, which ever comes first. Newer antifreeze formulations will last for 5 years or 150,000 miles. These antifreezes are usually red in color and are referred to as "Extended Life" or "Long Life" antifreeze. GM has been using this type of coolant in all their vehicles since 1996, called "Dex-Cool".

You can convert to the new long life coolant, but only if you completely flush out all of the old antifreeze. If any green coolant is allowed to mix with the red coolant, you must revert to the shorter replacement cycle.

Look for a shop that can reverse-flush the cooling system. This requires special equipment and the removal of the thermostat in order to do the job properly. This type of flush is especially important if the old coolant looks brown or has scale or debris floating in it. Always replace the thermostat with a new one of the proper temperature. It is cheap insurance.

The National Automotive Radiator Service Association (NARSA) recommends that motorists have a seven-point preventative cooling system maintenance check at least once every two years. The seven-point program is designed to identify any areas that need attention. It consists of:

a visual inspection of all cooling system components, including belts and hoses

a radiator pressure cap test to check for the recommended system pressure level

a thermostat check for proper opening and closing

a pressure test to identify any external leaks to the cooling system parts; including the radiator, water pump, engine coolant passages, radiator and heater hoses and heater core

an internal leak test to check for combustion gas leakage into the cooling system

an engine fan test for proper operation

a system power flush and refill with car manufacturer's recommended concentration of coolant

A Note on MGTD Overheating Problems

Many TD owners, maybe a third of us, complain that their car runs hot, often causing the gas to vaporize in the fuel lines, causing rough running. I've never had such a problem, even when driving mine in a parade, so I've given some thought to why this happens.

I think there are three possible causes. First might be that deposits have built up in the radiator. Such deposits slow the coolant flow through the small metal tubes, plus the deposits themselves act as insulation, slowing the transfer of heat through the metal. This is the first thing to suspect, but the problem still may persist after having the radiator professionally cleaned.

The second may be similar deposits building up in the engine. When this happens, the engine temperature gauge may appear normal (remember, your gauge is reading the temperature of the coolant, not the engine block) but the problem is that the heat isn't being transferred from the metal block to the coolant. Instead, the heat is radiating directly off the engine and heating the fuel in the fuel lines. Again, a well-equipped shop can flush the entire cooling system which may remove enough of the deposits.

The third problem might not be repairable. Recall that in manufacturing the engine block, the internal passages are made by making models of the passages using sand and wax and inserting them in the proper locations within the mold. The liquid metal melts the wax and the remaining sand is purged from the channels after the casting cools. I suspect that the casting process was crude enough back then that the channels didn't always form properly, or that the roughness of the channel walls varied between blocks, but were good enough to pass the limited engine tests done at the factory. In any engine, normal hot spots can cause the coolant to vaporize into steam, but these bubbles will be carried downstream and eventually condense out. However, if enough of these bubbles get trapped in small corners or along rough channel walls, steam pockets form. Heat does not transfer to the steam as well as the coolant, so these hot spots do not get cooled properly and the heat radiates outward into the air. Products such as Water Wetter work to break down these steam pockets and thus help engine cooling. People who've tried it tell me it lowers the temperature by 10 to 15 degrees.

Editor's note – I wrote this article based on an article by Charles Ofria at http://www.familycar.com/classroom/coolingsystem.htm, supplemented by information I found at http://auto.howstuffworks.com/cooling-system2.htm, and from my own experience.

Almost Everything You Wanted to Know About Two-Cycle Engines

Knowing that it takes four processes, and one stroke for each, to make a four-cycle internal combustion engine work, I always wondered how a two-cycle engine could do it in half the motions. I also always wondered why oil has to be mixed with gas. Here's what I found out.

Two-stroke engines have three important advantages over four-stroke engines:

- Two-stroke engines do not have valves, which simplifies their construction and lowers their weight and cost.
- Two-stroke engines fire once every revolution, while four-stroke engines fire once every other revolution. This gives two-stroke engines a significant power advantage over a four-cycle engine of the same weight.
- Two-stroke engines can work in any orientation, which can be important in something like a chainsaw. A standard four-stroke engine may have problems with oil flow unless it is upright, and solving this problem can add complexity to the engine.

Two-cycle engines are less expensive to build compared to four-cycle engines. They are lighter in weight and they also can produce a higher power-to-weight ratio. For these reasons, two-cycle engines are very useful in applications such as chainsaws, string trimmers, outboards, lawnmowers and motorcycles, to name just a few. Two-cycle engines also are easier to start in cold temperatures. Part of this may be due to their design and the lack of an oil sump. This is a reason why these engines also are commonly used in snowmobiles.

This is a cross section of a typical two-stroke engine:

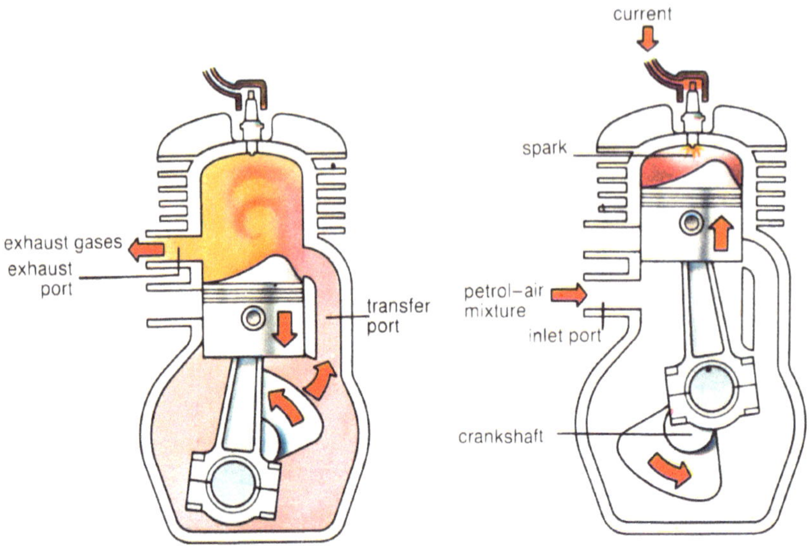

To understand the cycle, start with the point where the spark plug fires. Fuel and air in the cylinder have been compressed, and when the spark plug fires the mixture ignites. The resulting explosion drives the piston downward. As the piston moves downward, it compresses the air/fuel mixture in the crankcase. Then, as the piston approaches the bottom of its stroke, the exhaust port is uncovered (as opposed to a valve opening in a four-cycle engine). The pressure in the combustion chamber drives most of the exhaust gases out, as shown above. Also at this point, the transfer port opens so that the compressed mixtue in the crankcase can flow into the combustion chamber and help force out the exhaust. Note that in many two-stroke engines, the piston is shaped so that the incoming fuel mixture is deflected upward, so it doesn't simply flow right over the top of the piston and out the exhaust port.

Now the momentum in the crankshaft starts driving the piston back toward the spark plug for the compression stroke. While the air/fuel mixture in the combustion chamber is compressed, a partial vacuum is created in the crankcase. This vacuum sucks air/fuel in from the carburetor through the inlet port to be used for the next power stroke. Once the piston makes it to the end of the compression stroke, the spark plug fires again to repeat the cycle..

As you can see, the piston really is doing three different things in a two-stroke engine:

- On one side of the piston is the combustion chamber, where the piston is compressing the air/fuel mixture and capturing the energy released by the ignition of the fuel.
- On the other side of the piston is the crankcase, where the piston is creating a vacuum to suck in air/fuel from the carburetor and then pressurizing the crankcase so that air/fuel subsequently can be forced into the combustion chamber.
- Meanwhile, the sides of the piston are acting like valves, covering and uncovering the intake, exhaust, and transfer ports formed into the side of the cylinder wall.

This double function of the piston and crankcase is what makes two-stroke engines so simple and lightweight.

So, why is a two-cycle engine more powerful? The simple answer is that it completes a power cycle in only one crankshaft revolution, or one up and one down movement, as compared to a four-stroke engine, which uses four strokes, or two revolutions. This is accomplished by the end of the combustion stroke and the beginning of the compression stroke happening simultaneously and performing the intake and exhaust functions at the same time. The key to making this work is using the volume below the piston (where the oil reservoir resides in a four-cycle engine) as a charging pump to compress air for the next combustion stroke. All this makes for the two-stroke engine to have a greatly reduced number of moving parts, be more compact and significantly lighter. Two-stroke engines can provide a high power-to-weight ratio, but usually in a narrow range of rotational speeds.

You don't normally see two-stroke engines in cars, however. That's because two-stroke engines have some significant disadvantages.

If you ever have used a two-stroke engine, you know that you have to mix special two-stroke oil with the gasoline. In a four-stroke engine, the crankcase is completely separate from the combustion chamber, so it is filled with oil to lubricate all moving parts. In a two-stroke engine, with the crankcase serving as a pressurization chamber to force air/fuel into the cylinder, there is no place for an oil reservoir. Instead, oil must be mixed with the gas to lubricate the crankshaft, connecting rod and cylinder walls.

Here are more disadvantages:

- Two-stroke engines don't last nearly as long as four-stroke engines. The lack of a dedicated lubrication system means that the parts of a two-stroke engine wear faster.
- Two-stroke oil is expensive, and it takes about 4 ounces of per gallon of gas. In automotive use, a two-cycle engine would burn about a gallon of oil every 1,000 miles.
- Two-stroke engines do not use fuel efficiently and would deliver fewer miles per gallon.
- Two-stroke engines produce more pollution, which comes from two sources. The first is the combustion of the oil. The oil makes all two-stroke engines smoky to some extent, and a badly worn two-stroke engine can emit huge clouds of oily smoke. For the second reason, look again at the figure above. Each time a new charge of air/fuel is loaded into the combustion chamber, part of it leaks out through the exhaust port, which still is partially open. That's why you often see a sheen of oil around any two-stroke engine.

Because two-cycle engines double the number of power strokes per unit time when compared to four-cycle engines, power output is increased. However, it does not increase by a factor of two, as might be expected. The outputs of two-cycle engines range from only 20 to 60 percent above those of equivalent-size four-cycle units. This is because up to 30 percent of the unburned fuel/oil mixture escapes through the exhaust port before it closes. In addition, the exhaust gas isn't completely purged and a portion remains in the combustion chamber during the subsequent compression cycle.

As already discussed, lubricating traditional two-cycle engines is done by mixing the oil with the fuel. As expected, the oil is burned upon combustion of the air/fuel mixture. Direct injection two-cycle engines, making a comeback in motorcycles and snowmobiles, are different because the fuel is injected directly into the combustion chamber while the oil is injected directly into the crankcase. This process is efficient because the fuel is injected after the exhaust port closes, and therefore more complete fuel combustion occurs and more power is developed. Direct injection engines have a higher power-to-weight ratio than traditional two-cycle engines. Because the oil is directly injected into the crankcase, less oil is necessary and lower oil consumption results.

Thanks to the following sources for the above wealth of information:

1. http://science.howstuffworks.com/transport/engines-equipment/two-stroke7.htm
2. http://www.amsoil.com/articlespr/article_2cycleapplications.aspx
3. http://en.wikipedia.org/wiki/Two-stroke_engine

Thermal Efficiency

The modern cars that most of us drive every day are quite a bit different from the classic/antique cars we drive for fun. However, under the bonnet, all these cars share one common problem. Gasoline engines just are not that efficient in converting the energy stored in gasoline into mechanical energy to propel the car. That doesn't mean they don't run well or give enough power. It just means they can use only a small proportion of the gasoline to actually move the car. It's not the engines fault. Despite major improvements in engine efficiency in the last five decades or so, there's a brick wall the designers are pushing against. Significant, and often expensive, "tweaks" in how the engine operates have been incorporated into the engines to produce even miniscule mileage improvements.

We all understand that the engine has to burn a mixture of fuel and air in the cylinders. Keep in mind that any heat exiting the engine represents wasted energy. But, before all that energy from the burning fuel can be utilized, a large amount of it exits the exhaust pipe, and a lot of heat goes out with it. At the same time, more heat is escaping through the cylinder walls into the coolant, which then gets pumped to the radiator to prevent the engine from overheating. When we account for these two major energy losses, only about a quarter of the fuel energy actually reaches the flywheel. But, from the flywheel we still have to get the power to the tires. More energy is lost to friction in the gearbox, more in the drive shaft bearings and differential, and still more in the flexing of the tires as they grab the road. That's why our tires get warm. Altogether, only about 16% of the energy in the fuel gets used to move the car. In other words, our engines first convert all the fuel that gets burned into heat (and not all of the fuel even gets burned), then throw 84% of it away and use what's left to propel the car.

So, where does it go? The wasted 84% has been turned into exhaust gas or hot air which radiates outward and quickly is absorbed into the atmosphere. The last 16% is mostly used to overcome aerodynamic drag, pushing air out of the way as we drive. This swirling air is turbulent with energy supplied by the engine. No doubt you've noticed the wind caused by a passing car or truck! This turbulence dissipates as soon as the car passes, being converted into more heat, which is also absorbed into the surrounding air. So, after we get home from a drive and our car has cooled off, all the energy from the fuel we used on the journey has been turned into heat and released into the atmosphere (except for the energy used to put a dent into someone else's car or to push a chunk of concrete curbing a few inches).

There's not much we can do about this. It's one of the basic laws of thermodynamics we have to live with. The world still needs motorized transport. Modern auto companies have tried hard to improve things, like making cars lighter and more aerodynamic. They also use electronic fuel injection and ignition to improve combustion and timing, but still they only can push the laws of physics, not break them. And, many of these little improvements offer only a slight improvement at an increased production cost and give you something else to break on your car. Others claim their clean gas/electric hybrids get increased mileage, but conveniently forget about the electricity sometimes needed to recharge them. They don't mention they are really switching some of the energy we consume away from our gas tanks to a power station many miles away.

If we think our cars are wasteful, how about the old railroad steam locomotives. The best ones, on a good day were even less fuel efficient than our automobiles. Then why is rail transport considered the second most efficient method of transporting heavy loads? That's because railroad cars have solid steel wheels running on steel tracks. At the point of contact there is almost no distortion of the rails or wheels and that leads to very low rolling resistance. It's what enabled steam locomotives to pull loads more than a hundred times their own weight at speeds up to seventy miles per hour. These days modern railway locomotives use high (20:1) compression diesel engines and have thermal efficiencies up to 30%, which explains why steam engines were replace by diesels.

What is the most efficient mode of transportation ever? Believe it or not, it's the canal barge. In the old days we had a good network of canals and they all had a towpath running alongside. It used to be a common sight to see a heavily-loaded barge gliding through the water at about two or three miles per hour with a long tow rope attached to a horse or two. Some of those barges were carrying up to forty tons of goods and it was all being moved by just one or two horsepower. Slow maybe, but extremely efficient, and the only fuel needed was a bucket of oats. These days it's still an amazing sight to see the long barge trains on the Mississippi River, all tightly tied together and being pushed from behind by a single powerful tug. They are carrying thousands of tons of cargo and being moved along by just one tugboat.

And what about those jumbo jets? A Boeing 747 cruising at almost seven miles high at 540 miles per hour burns about one gallon of fuel every second, which calculates to 0.15 miles per gallon. However, as it is carrying about three hundred passengers, this is equivalent to forty-five passenger-miles per gallon.. A large SUV or van could just about match it if they carried five or more people, but would take a lot longer to cross the country.

And the biggest fuel guzzler? It had to be the Saturn V rocket used to carry men to the moon in the late 60s/early 70s. The first stage lifted the 3,000 ton vehicle off the pad and accelerated it to 5,000 mph, consuming fuel and liquid oxygen at the rate of fifteen tons per second. To supply this amount of fuel/oxygen to the engines, the fuel pumps were using about 55,000 horsepower. The Saturn V first stage not only set a new record for fuel consumption, it also broke records for noise volume. Measurements made near the launch pad at liftoff showed it made the loudest man-made noise except for a nuclear explosion. Spectators over two miles away said the noise was so intense that they felt physical pain. That's not surprising, because the Saturn rocket's noise spectrum was most intense in the lower frequencies where it could cause resonances in some of the larger bones in our bodies.

The first person to investigate the efficiency of heat engines was Sadi Carnot, a French scientist in 1819. While serving in Napoleon's army, he'd seen a large British steam engine working and became curious about its thermal efficiency. He concluded that the efficiency was determined only by the temperature difference between the hot inlet steam and the cooler exhaust steam; the greater the temperature difference, the higher the efficiency (i.e., more heat energy was being extracted from the steam and converted into mechanical energy). Even though the nature of heat was not fully understood then, his reasoning was sound and the theory still holds true today. It applies to every type of heat engine, including gasoline engines, diesels, jet engines and even the steam turbine section of nuclear power stations. When a method to perform a mathematical efficiency analysis was developed, it was determined that the efficiency of the big British steam engine was not quite 2%.

This write up was based on an article by Peter Bland that appeared in British Cars of New Hampshire.

Author's Note – It also takes energy to produce noise, run air conditioning, lights, etc. That energy has to come from the fuel that is burned. And, there are inefficiencies within the engine. While most of the exhaust is water vapor and carbon dioxide, other pollutants are the result of incomplete combustion. This takes place near the cylinder walls, which are cooler and thus inhibit complete combustion of the fuel and air. A professor of mine back in college designed an engine that injected air through the cylinder walls to keep the fuel in the middle of the cylinder and thus produce a more complete combustion. The theory was sound, but he never could sell the idea to the auto industry. Another potential design is the ceramic engine. Because ceramics do not melt and barely expand with temperature change, they don't need to be cooled. Operating at higher temperatures could produce greater thermal efficiency, but how would it be lubricated and how would you keep the paint from blistering on the bonnet?

Automotive Air Conditioning

As a little kid growing up in the 50s, I guess I figured everyone just suffered in the summer heat. Well, maybe we did, but that didn't mean that anyone enjoyed it. The desire to keep cool during the hot summer months goes back thousands of years. Probably the oldest form of "air conditioning" was a simple hand-held fan. I suppose if you were wealthy enough, one could hire someone to wave a larger fan to blow hot, ambient air at you and let the evaporation of sweat from your skin cool you off just a bit.

For millennia (probably), people harvested ice in the winter from frozen lakes and stored it underground to preserve food for as long as possible until the last of the ice melted. When someone was suffering from a high fever, that ice could be packed around them to help lower their temperature and bring them some relief. But naturally-formed ice had its problems as people crowded into ever-growing cities: the harvested ice often came from lakes and rivers that were contaminated with sewage, causing health concerns. That was the extent of getting relief from the heat until the early 1900s.

A Scottish scientist named William Cullen discovered the principle of artificial refrigeration in the 1740s. He heated a liquid into a gas, and then observing that as the gas condensed back into its liquid form, it was cold. Cullen never turned his theory into any sort of working device, but many others picked up on his idea. This is the principle behind refrigeration that we still use today. Benjamin Franklin even conducted experiments in 1758 by evaporating alcohol to attain freezing temperatures, but this process certainly would have been highly inefficient, as first one would have to make the alcohol.

In 1834, American inventor Jacob Perkins, living in London at the time, built the world's first working vapor-compression refrigeration system, using ether in a closed cycle. His prototype system worked and was the first step to modern refrigerators, but it didn't succeed commercially.

At this point, let me explain briefly how the principle of air conditioning works. There are two main components, the compressor (sometimes called a condenser) and the evaporator. When a gas is compressed, it gets hot. This is because energy is being forced into it. When compressed to the point that it liquefies, it gets very hot. Then, the hot liquefied gas is run through metal coils and air is blown over them to dissipate the heat and thus cool the liquid (feel the air blowing off the compressor unit outside your house, and you'll notice it is warm). Once the hot liquid cools off, it is pumped into the house and allowed to expand in the evaporator (located inside your furnace) back into a gas. In doing this, the gas now gets very cold. Blow warm inside air over the metal tubes containing the cold gas and the air now gets cold. As an interesting side note, take a thick rubber band and stretch it out as tight as you can, then touch it to your chin or cheek. You'll notice it is hot. Hold it tight for a moment to let it cool off, then let it relax completely and feel it again. It's cold! Same principle, but I doubt a rubber band based air conditioning system ever would catch on.

The first modern air conditioner was invented in 1902 by Willis Carrier, but he wasn't really looking for a way to cool air per se, but to dehumidify it. A skilled engineer, Carrier was looking for ways to control humidity at a printing plant in Brooklyn, NY. Using the concepts of mechanical refrigeration established years earlier, Carrier's system sent air through coils filled with cold water, removing moisture to control room humidity, while at the same time cooling the air. Humidity control actually is a by-product of cooling the air. The warmer the air, the more moisture it can hold. Cold air can hold less moisture. Thus, when humid air is cooled, excess water vapor condenses out in the form of liquid water. Carrier obviously realized the potential for using his device primarily to cool air, and eventually formed his own company in 1933, the Carrier Air Conditioning Company of America. This became the model in the growing U.S. marketplace for air-cooling systems.

While air conditioning became popular in the 1930s, it was expensive and most often found only in large stores and theaters. Storefronts often advertised that they were air conditioned to attract shoppers. Home units were available but they were expensive.

At this point, I'll divert from home/industrial/commercial air conditioning and look at its development for automobiles. By now, the technology that allowed air conditioning was well-developed. Aside from being adapted for use in cars, it pretty-much was the same.

There was no need for air conditioning the first automobile when it was patented in 1886, as it was little more than a motorized tricycle. An umbrella could be attached to the frame to give some relief from the sun. The Kool Kooshion seat cover was introduced in 1919. This device used small springs to hold drivers about a half-inch above the car seat, allowing air to circulate behind and beneath them. It essentially allowed sweat to evaporate to help the driver keep cool. Kool Kooshions are still sold today at many major retailers!

By 1921, most cars were enclosed. The Knapp Limo-Sedan Fan was a small electric fan that could be mounted inside of a car to make driving in the heat more tolerable. Nine years later, in 1930, the "Thermador Car Cooler" came along. It used the evaporation of water (rather than one's own sweat) to cool air. The cooler air then was blown into the car through the open passenger-side window. Although it was the first device to actually lower the auto cabin air temperature, it only worked in areas with very low humidity, besides being a bit awkward looking. Anyone who has lived in a desert climate certainly is familiar with the "swamp cooler." This essentially was an automotive version of it. Vintage fans and coolers still can be found on internet auction sites.

In 1939, Packard became the first car manufacturer to offer air conditioning as an option, but it was a bit awkward to use. These rear-trunk unit required the user to install or remove the drive belt from the compressor to turn the system on or off. And, there were the only two settings: on and off. The option cost a whopping $274 at a time when the average yearly income was $1,368. Between 1940 and 1942, Packard equipped 1,500 automobiles with air conditioning. Most of the units were installed on expensive luxury cars sold in the Southwest. The price, combined with the start of World War II, caused the option to be impractical and short-lived.

By 1947, independent manufacturers began offering after-market air conditioners that could be installed on just about any make of car, creating a large business. By 1953, auto manufacturers again began offering air conditioning. Again, it was an option, all of them being rear-mounted systems, not much different from what Packard first offered in 1939.

Typical after-market air conditioner installed on an early Ford Mustang.

In 1954, Pontiac and Nash become the first two companies to fit the air conditioning system in the front of their cars instead of in the trunk. The Nash system combined the heater and air conditioner into one in-dash system, establishing the standard that most cars have followed to this day. But, there still were a lot of issues to be refined as manufacturers experimented with improvements like air circulation, placement of air outlets and issues regarding drainage for condensation. With the evaporator mounted inside the passenger compartment, sometimes the condensed water dripped over the passengers. This problem persisted into the 1950s as exemplified by a highly publicized incident when Mamie Eisenhower's dress was stained by the dripping condensation in an air conditioned Cadillac. In 1957, the Cadillac Eldorado Brougham became the first model to include air conditioning as a standard item.

The number of air conditioned cars nearly tripled from 1961 until 1964, and air conditioning started to be offered for trucks. With the big three's introduction of compact cars in the 1960 model year, air conditioning was not offered even as an option, but Corvair introduced air conditioning to the compact class on its 1961 models. In 1963, Corvette became the first sports car to feature air conditioning. In 1964, Cadillac introduced "Comfort Control." Drivers now could set a preferred temperature, and the system would automatically adjust the air conditioning or heater output to keep the car interior at that constant temperature.

Today, it is hard to find a car in the United States without factory-installed air conditioning. There are a few, but not many. Except for replacing R-12 refrigerant with R-134a by 1994 because of environmental concerns, automotive air conditioning systems haven't changed much except for system tweaks to improve efficiency, decrease weight, and make them quieter and more reliable. And now, even R-134a has been found to have some negative environmental impacts and is being replaced with R-1234yf, which has shown to have a minimal impact on the environment (*Editor's note...ya, right*). Maybe there ***is*** potential for a rubber band based air conditioner.

Thanks to the following sources:

1. https://www.materials.sandvik/en/campaigns/fridge-of-the-future/the-history-of-the-refrigerator/
2. https://www.asme.org/topics-resources/content/global-cooling-the-history-of-air-conditioning
3. https://fjcinc.com/the-history-of-automotive-air-conditioning/
4. https://www.curbsideclassic.com/automotive-histories/cold-comfort-history-of-automotive-air-conditioning-part-1-pre-world-war-ii/
5. https://www.airah.org.au/Content_Files/HVACRNation/2008/June2008/HVACRNation_2008-06-F02.pdf

The Development of Automotive Brakes

We press on the gas pedal to make our car go, then step on the brake pedal to make it slow down or stop. Besides that, we probably don't give much thought to our brakes as long as they are working properly. You probably give even less thought to the many incremental steps that were taken and the technologies developed to make brakes so nice that we hardly give their complexity a thought any more.

As the car accelerates, the chemical energy stored in the gasoline (or battery for you EV enthusiasts) is converted into kinetic energy, the type of energy associated with a mass in motion. To slow down or stop, that energy has to be dissipated. That's what the brakes do; they convert the kinetic energy of the car in motion into heat energy, which then is dissipated into the air or other adjacent metal parts. As car weight grew and typical driving speeds increased, the kinetic energy that the brakes had to dissipate also increased, putting more and more of a demand on them.

Here are the various braking systems, in approximate chronological order, that have evolved over the many years since the invention of the wheel.

Wooden block brakes – The earliest brake system consisted of nothing more than a simple lever that moved a block of wood against the steel-rimed wheel. This worked well for who knows how many years, as nothing moved much faster than a horse-drawn carriage or wagon. This even worked for early automobiles, which really were nothing more than motorized carriages, as they only hit speeds of 10-20 mph and in sparse traffic. But when rubber tires were introduced by the late 1890s, the steel-rimmed wheels became obsolete, as well as the wooden block brakes, as the rubber was worn away quickly.

External mechanical drum brakes – Drum brakes work by a shoe generating friction as it rubs against the surface of a brake drum attached to the wheel. Early brakes used a metal band surrounding the drum. The earliest mechanical drum brake was proposed by Gottlieb Daimler in 1899. It consisted of a cable wrapped around a metal drum attached to the wheel. In 1901, Wilhelm Maybach designed the first Mercedes with such a system on the rear wheels and it was operated by a hand lever. Sometimes, when drivers would go upward on hills, the brakes tended to slip. Also, because these brakes were exposed to dust and water, they would wear quickly and had to be replaced often. Early external drum brakes didn't last long and the technology was obsolete almost as soon as it went into service.

Expanding internal shoe brakes –Louis Renault was credited, in 1902, with inventing the drum brake as we know it today, placing expanding shoes inside the drum so that dust, water, and other elements would be kept out, or at least minimized. He used woven asbestos for the drum lining, as it dissipated heat well. Early drum brakes typically lasted for about 1,000 miles.

Four-wheel brakes – As vehicle speeds increased, manufacturers began to look for improvements in brakes. One of the earliest innovations was in 1903 when four-wheel brakes were fitted to a Dutch Spyker 60/80 hp model. A patent for the brake was granted in 1910 to Giustino Cattaneo of Isotta Fraschini, an early Italian auto manufacturer. Another development arose in the US in 1915 when Duesenberg created cars that applied internal brakes to both front and rear wheels, as their cars could reach 80 mph on a straightaway. In 1919, the Spanish automobile company Hispano-Suiza used a single foot pedal to operate four-wheel brakes in its H6B model. Before that, it was common to operate a separate hand and foot brake simultaneously. At the New York Auto Show in 1923, only Duesenberg and Rickenbacker had cars with four-wheel brakes, but by the following year, the number had increased to 26.

Hydraulic brakes – In 1918, Malcolm Loughead (who later changed the spelling of his name to Lockheed in 1926) proposed a concept of a four-wheel brake system using hydraulics. Using cylinders and tubes, Lockheed used fluids to transfer force to the brake shoe when the driver pressed a pedal. It required much less effort for the driver to apply brakes. The hydraulic brake system was first fitted into all four wheels of a Duesenberg Model A in 1921. The system suffered from fluid leakage issues, but engineers from the Maxwell Motor Corporation developed rubber cup seals to help solve the problem. In 1923, the improved Lougheed brakes were offered as an optional upgrade on the Maxwell-Chalmers car for $75. This new brake design was also used in Chrysler cars from 1924 to 1962. Other car manufacturers eventually followed Chrysler. By 1931, U.S. manufacturers such as Dodge, DeSoto, REO, Franklin Graham and Plymouth switched to hydraulic brakes. But Ford and General Motors stuck with mechanical brakes for a while. By the mid-1930s, GM incorporated Bendix hydraulic brakes. Bendix eventually purchased Lockheed's hydraulic brake company, and GM began to use hydraulic brakes on all of their cars. Ford was the last manufacturer to adopt hydraulics. They used a mechanical brake applied to a drum inside the transmission and stuck with this concept until 1938.

Disc brakes – Disc brakes are superior to drum brakes because of their greater stopping capability and better ability to dissipate heat. The disc brake was invented decades before it became popular. In 1898, Elmer Ambrose Sperry designed an electric car with front-wheel disc brakes built by the Cleveland Machine Screw Co. Disc brakes work like bicycle hand brakes, wherein a caliper with brake pads pinches a disc or rotor. However, it was William Lanchester, an English engineer, who patented the idea in 1902. The biggest downside to his invention though was the horrible screeching noise it produced, which was caused by copper brake linings moving against a metal disc. After five years, another Brit named Herbert Frood solved the noise problem by lining the pads with asbestos, which continued to be used in car brakes until the 1980s. Still, disc brakes were not yet popular. They only began to be widely used in Europe during the 1950s when vehicle weight and speeds were increasing, causing drum brakes to become less efficient because they didn't dissipate heat as well. In 1949, Crosley Motors became the first American manufacturer to use disc brakes. They were fitted to Crosley's Hotshot model but discontinued in 1950. These brakes, built by Auto Specialists Manufacturing Company (Ausco), used twin discs that spread apart and rubbed against the interior of a cast-iron drum. Less pedal pressure was required than with caliper discs, and they provided more friction surface than drum brakes. In 1962, Bendix four-wheel disc brakes were standard on the high-performance Studebaker Advanti and as optional extras for Hawk and V8 Lark models. It took only a few years for other manufacturers to adopt disc brakes since drum brakes were proving inadequate for the increasing speed and size of cars being built. During the 1960s, many auto manufacturers worldwide started switching to disc brakes. Some of the companies that were the first to do so in their countries were Italy's Lancia in 1960, Germany's Mercedes-Benz in 1961, France's Renault in 1962, Japan's Nissan in 1965 and Sweden's Volvo in 1966.

Power-assisted brakes – The power brake, which reduces the amount of force the driver needs to apply to the brake pedal, is a standard feature of most cars today. But, it actually was one of the earliest innovations with brakes – Chicago-based automobile maker Tincher first made brake assist available in 1903, using a small air pump to assist in stopping the car. In 1928, the Pierce-Arrow featured a vacuum-operated power booster for brakes. This vacuum-assisted brake booster was designed originally for aircraft. The intake manifold supplied the vacuum needed to reduce the amount of effort needed to apply the brakes. The vacuum booster brake system we know today has its origins in 1927 when Belgian engineer Albert Dewandre invented the servo-brake or brake booster system he called "Dewandre." In the same year, Chandler cars came with a Westinghouse Vacuum Booster. By the 1930s, vacuum-assisted drum brakes were fitted into Cadillac, Lincoln, Mercedes, Duesenberg, and Stutz cars. While the vacuum boost power brakes were the most popular type of power-assisted brakes, there were other power brake systems that began to appear. The Hydrovac system started to be available in the 1940s. As the driver applied pressure to the brake pedal, the fluid pressure was increased by a slave cylinder in addition to the primary wheel cylinders. In the 1950s, Bendix introduced the Treadle Vac and was available on all GM cars during the 50s. It soon also was available on Lincoln, Mercedes, Hudson, Nash, Mercury and Edsel models. However, the Treadle-Vac was a single line system, meaning a failure on any hose or joint could impair the entire system.

Self-adjusting brakes – As the brake components gradually wore away, the gap between the surfaces needed to be adjusted to minimize the amount of pedal travel. In 1925, Cole Motor introduced one of the earliest self-adjusting brakes. They fitted it on the Series 890 Cole during their last year of production. Simultaneously, Jowett Cars also installed their self-adjusting brakes to all four wheels of their Sedan, Brougham and Touring models as an option. After the demise of these two brands, self-adjusting brakes did not appear again until 1946 when Studebaker used a Wagner Electric mechanism. In its self-adjusting feature, as the linings wore down, a simple mechanism kept them at the same distance from the drums. This type of brake was included in the 1957 Mercury, in the 1958 Edsel, and AMC cars in the mid-1960s.

Anti-lock brakes – No doubt you were startled a bit the first time you ever experienced the clanging sound and thumping of anti-lock brakes kicking in. The anti-lock (sometimes called anti-skid) brake system, or ABS, was created to keep the driver in control while braking hard, by keeping the brakes from locking up. Anti-lock brakes were introduced by French engineer and aeronautical pioneer Gabriel Voisin in 1929 for use in airplanes. The system uses speed sensors on each wheel and constantly compares the speed of the four tires. One that is about to lock up will turn slower than the others since it is slipping on slick pavement. The system then activates hydraulic valves to reduce the pressure of brake fluid on that single wheel, keeping it rolling, and helping the car to continue traveling straight and reducing the chance of the car skidding or spinning. While stopping distance may increase, there is an overall safety advantage in that the driver can remain in control and still steer the car. (You do maintain extra stopping distance while driving on wet or icy pavement, don't you?) In 1958, the Road Research Laboratory (RRL) and Dunlop developed a practical, mechanical ABS for a car and tested it on a Jaguar Mark VII fitted with disc brakes. In 1966, it was fitted in a production car, the Jensen FF sports sedan in Great Britain. Meanwhile, in the US, Ford offered an anti-skid system as an option for the 1964 Lincoln Continental Mark II. However, it was very expensive to produce, and soon was withdrawn. In 1968, Ford introduced the "Sure-Track" analog anti-lock brake system which operated only on rear wheels. Production costs still were too high, so it was initially offered as an option for the Thunderbird until it became standard on the 1970 Mark III. Other companies soon followed suit on most cars.

Automatic brakes – While self-driving cars probably still are a way off, many of the technologies developed for them have found their way into production cars. In 2006, Mercedes introduced the Brake Distronic Plus system, which uses radar to sense the distance and closing rate to a vehicle ahead and can bring a car to a stop even if the driver does not touch the brake pedal to prevent a collision. This technology is rapidly advancing and can be found as standard on many higher-end models and as an option on many lower end models. As the push continues towards self driving cars, intelligent automatic braking surely will become a common technology.

Regenerative Brakes – As I mentioned earlier, a moving car has kinetic energy that was provided either by the burning of a fuel or depletion of a battery, and the brakes dissipate that energy to bring the car to a stop. On gasoline powered vehicles, that energy is converted to heat by the friction of brake components rubbing together and is released into the air. But, on hybrid or electric vehicles, much of that energy can be recovered. A motor and generator basically are the same device: a motor converts electrical into mechanical energy, while a generator converts mechanical into electrical energy. With regenerative braking, as the driver steps on the brake, the motor works in reverse. Instead of using friction to slow the car, the hybrid or electric vehicle uses its motor as a generator, using the electricity generated to partially recharge the battery. I would assume that at some point the mechanical brakes are applied, depending on how hard of a stop the driver commands, but much of the braking still is done with the regenerative capability. Energy recovery can range from 16 – 70%, dependent largely on the drivers' driving habits.

Sources:

1. https://didyouknowcars.com/the-history-of-brakes/
2. http://www.edubilla.com/invention/drum-brake/
3. https://driving.ca/column/how-it-works/how-it-works-regenerative-braking
4. https://4lifetimelines.com/blogs/knowledge-garage/the-history-of-automotive-brake-systems
5. http://greeninginc.com/blog/uncategorized/a-cumulative-history-of-brake-technology/

A Brief History of Automobile Tires

Who knows when the wheel was invented. After centuries (or milenia) of placing heavy objects on several logs placed side by side, someone obviously discovered that they could slice off a piece of the log and mount it on a shaft. Presto...no more moving the log from behind the object being moved and repositioning the log in front. After a few miles, I suppose the wheel would split and they'd have to make a new one. But, still, it was an improvement.

The earliest tires were bands of iron (later steel), placed around the wooden wheel. The tire would be heated enough to make it expand, then placed over the wheel and quenched quickly, causing the metal to contract and fit tightly on the wheel. The outer ring served to "tie" the wheel segments together for use, and also provided a wear-resistant surface on the perimeter of the wheel. The word "tire" (or "tyre") thus emerged as a variant spelling to refer to the metal bands used to "tie" the wheels. This work was carried out by a skilled worker known as a wheelwright.

The qualities of early rubber made it impractical to use for most industrial applications. At high temperatures, rubber would melt or at least become sticky. At low temperatures, it would become brittle. One of the early discoveries that made rubber an excellent material for tires was the process called vulcanizing. During vulcanization, the milky latex from a rubber tree is combined with a curing ingredient — usually sulphur — and heated under pressure. This process, which was discovered by an American inventor named Charles Goodyear in 1839, made the rubber more stable, tougher and more resistant to heat, which made it more useful for industrial purposes. Its name for this process comes from Vulcan, the Roman god of smiths.

Goodyear had experimented with rubber for several years before discovering vulcanization. A former hardware store owner who had no scientific training or knowledge of chemicals, it is often said that his discovery was by accident after a mixture of sulphur and rubber gum splattered onto a hot stove and formed a hardened material.

Goodyear, however, denied that it was accidental, but rather the result of a series of experiments and observations.

Goodyear sent samples of his cured rubber to British rubber plants. An Englishman named Thomas Hancock, who had been trying to make weatherproof rubber for 20 years, saw one of the samples and he noticed the yellowish powdery residue of sulphur on the surface of the sample. Hancock reinvented the process in 1843, four years after Goodyear. Goodyear obtained the United States patent for this process in 1844, but when he applied for the British patent, he found that Hancock had beaten him to it.

The first solid rubber tires appeared in the mid-1800s. Pneumatic, or air-filled, tire, was patented as early as 1845. Despite the pneumatic tires better ability to absorb shock, solid rubber tires still were preferred because of their durability, so pneumatic tires never really caught on. The popularity of bicycles in the late 1800s revived the idea of the pneumatic tire, and in 1888 a veterinary surgeon in Belfast, Ireland named John Boyd Dunlop obtained a patent for a pneumatic bicycle tire. Necessity being the mother of invention, Dunlop did this in an effort to prevent the headaches his son would get while riding on rough roads.

The first use of pneumatic tires for automobiles was pioneered by the Michelin brothers, André and Édouard. They equipped a car with pneumatic tires and drove it in the 1895 Paris-Bordeaux road race. Though André and Édouard didn't win the race, they generated popular interest in pneumatic tires, and Michelin & Cie became a leading producer of tires in Europe. Aided by legislation that discouraged their use because they were hard on the roads, solid rubber tires quickly disappeared from use.

In 1898 Goodyear Tire and Rubber Company—named after George Goodyear, the discoverer of vulcanized rubber—was formed in America by Frank Seiberling. Then, the Firestone Tire & Rubber Company was started by Harvey Firestone in 1900. Other tire makers followed.

For the next fifty years, automobile tires were made of an inner tube that contained compressed air and an outer casing that protected the inner tube and provided traction. The rubber that made the casing was reinforced by layers or "plys" of rubberized fabric cords embedded in the rubber. The tires made during this period were known as bias-ply tires because the plys ran across the tire in alternating diagonal layers at about a 55 degree angle to the wheel rim. Bias-ply tires continue to be made and are sold as authentic equipment for antique and collector cars that were made during this period.

Michelin first introduced steel-belted radial tires in Europe in 1948. Radial tires are so named because the ply cords radiate at a 90 degree angle from the wheel rim, and the casing is strengthened by a belt of steel fabric that runs around the circumference of the tire. In radial tires the ply cords are made of nylon, rayon, or polyester. The advantages of radial tires include longer tread life, better steering characteristics, and less rolling resistance, which increases gas mileage. On the other hand, radials have a harder riding quality, and since they are technologically more complex than bias-ply tires, they are about 45 percent more expensive to make. Because of their construction, radial tires require a different suspension system from that used by cars designed for bias-ply tires. It is generally recommended that radial tires not be used on cars designed for bias-ply tires. (Editor's note – I have radial tires on my MG TD and was amazed by the more-solid feel they gave the car. But, I don't drive the car aggressively or at highways speeds; 55 is about my limit, and then just for short spurts.)

Over the next 20 years radial tires became standard on new cars outside of America. Michelin in France, Bridgestone in Japan, Pirelli in Italy, and Continental in Germany became major radial tire manufacturers. Automobile tires everywhere became tubeless as tire technology improved. But, American manufacturers were hesitant to switch from bias to radial tires.

Both the American automobile manufacturers and the tire companies fought the radial tire. Detroit, home of the American automobile, was afraid of how much it would cost to redesign automobile suspensions to accept radial tires. The tire industry was afraid of how much it would cost to retool the entire American tire industry to make the more costly radial tires. Not happy with the threat of having to make tremendous investments, most American automobile makers and tire manufacturers wrote off the radial tire as "a freak product that isn't going anywhere."

With the notable exception of B.F. Goodrich, American tire companies felt that the American public wasn't ready to pay a lot more for the harder ride that radials produced, and they stuck to making bias ply tires. Goodrich bucked this trend by investing heavily in radial tire technology, only to have their tire—the Silvertown Radial 900, introduced in 1965—snubbed by the American automobile industry. Eventually Goodrich sold its tire operations and got out of the tire business.

In 1967, Goodyear, the world's largest tire company, introduced their response to the radial, a bias-belt product called the Custom Superwide Polyglas. The bias-belted tire simply added a fibreglass belt to the bias-ply tire. The bias-belted tire would last 30,000 miles compared to 40,000 for radial and 23,000 for bias-ply tires. It could be used on cars designed for bias-ply tires. Best of all, it could be made on existing bias-ply tire-making machines, which made its cost not much more than a bias-ply tire.

Fueled by an advertising blitz, Goodyear's bias-belted tire sales rose from 2 percent of the original-equipment market in 1968 to 87 percent by the early 1970s. In advertisements touting their bias-belted tires, Goodyear ridiculed radials for their hard ride and their high cost. The American tire and automobile industry was confident that the bias-belted tire would keep the interest in the radial tire down for a while and expected to have plenty of time to develop their own radial tire expertise at their own pace.

Then came 1973 and the gasoline crisis. Gas went from 30 cents to a dollar a gallon. Americans demanded more economical cars. That year, imported cars represented 15% of American auto sales, but by the early 1980s imports were 28 percent. Of course, each foreign car came equipped with radial tires. Americans clamored for radial tires when they found that they improved gas mileage. Companies like Michelin and Bridgestone were only too happy to supply the American market.

In the mid-seventies, Firestone Tire decided to get into radials on the cheap, fabricating radial tires on machines made for building bias tires. The tires came apart in a spectacular manner. Firestone recalled close to 9 million of its Firestone 500 steel-belted radial tires. From 1977 to 1980, Firestone's tire business dropped 25 percent, resulting in financial losses, layoffs, and a drop in its stock price. Firestone was rescued when Bridgestone Tire bought them in 1988.

Goodyear finally produced a radial tire in 1977 by investing billions of dollars in radial technology. Other American tire companies either merged or were bought out. By 1983, all new American cars came with radial tires.

One last note about tires. Since natural rubber is white, why are tires black? It's because of an additive called carbon black. Carbon black is a byproduct of the combustion of various petroleum products. When added to rubber, it forms a chemical bond that increases abrasion resistance and tensile strength significantly – which helps lead to a long-wearing tire. A tire made without carbon black likely would only last 5000 miles or less.

Further, the carbon black helps to conduct heat away from the tread and belts of the tires, which also helps to increase the life of the tire. The carbon black compounds also help protect the tires from UV rays and ozone, which can shorten the lifespan of tires.

One further benefit of carbon black in tires is that it helps make rubber compounds more electrically conductive. Without carbon black in the tires creating a conductive path from the vehicle to the road, static charges may be more likely to build on vehicles as they travel and lead to an unanticipated static discharge, or electric shock.

And last, black tires are easier to keep clean. Ask any classic car owner – whitewalls are a pain to keep gleaming, as they show dirt and discoloration easily and can stain from contaminants on the road. Black tires, on the other hand, are easier to keep looking fresh.

Sources:

Wikipedia

1. http://www.jags.org/TechInfo/2001/05May01/tires/historyoftires.htm

2. http://auto.howstuffworks.com/how-can-i-make-my-tires-last-longer.htm
3. http://www.wisegeek.org/what-is-vulcanized-rubber.htm
4. https://www.goodyear.com/en-US/learn/tire-basics/why-are-tires-black

A Brief History of the Automotive Battery

We usually don't give our car's battery much thought until we turn the key (or push or pull the starter knob, as appropriate) one morning and hear nothing. Despite various improvements, the basic design of your car's lead-acid battery has remained almost unchanged for over a century.

Your car's battery supplies electrical current to the starter and other electrical components. Once the engine is running, the alternator takes over to provide power for the car's electrical systems and to recharge the battery. If the car's battery loses its charge, which it will do eventually, you'll probably hear nothing when trying to start it. While there are many types of batteries for numerous applications, cars have almost universally stuck with lead-acid batteries.

Automotive batteries are designed to deliver maximum current for a short period of time, as compared to your cell phone's battery, which provides a trickle for a very long time. Typically, starting the engine discharges less than three percent of the battery capacity, which the alternator quickly replaces once the engine is running. A full discharge can reduce the battery's lifespan, or be the "straw that broke the camel's back" if it already is weak.

Modern automobile batteries use six series-connected cells to provide a nominal 12 volt system (in most passenger vehicles and light trucks), or twelve cells for a 24 volt system in heavy trucks or earth-moving equipment. Despite being relatively large and heavy for their voltage output, lead-acid batteries are ideal for automotive use because they can provide a large amount of current for starting.

In addition to starting the engine, the battery supplies the extra power necessary when the vehicle's electrical requirements exceed the supply from the charging system (my TD typically will show the battery discharging when I drive with the headlights on). It also stabilizes the electrical flow, evening out potentially damaging voltage spikes. While the engine is running, the voltage regulator keeps the alternator's output between 13.5 and 14.5 V.

Early cars did not have batteries, as their electrical requirements were minimal: a bell or manual horn was used instead of an electric horn, headlights were gas-powered, and the engine was started with a hand crank. The Hudson Motor Car Company was the first to use a standardized battery in 1918. Car batteries quickly caught on and by 1920 many cars were equipped with electric starters. The sealed battery, which did not require refilling, was introduced in 1971 (certainly you remember the gas station attendant raising your hood, popping the caps on your car's battery to check the electrolyte level, as well as your oil and coolant levels…all while pumping your gas for you).

Early electrical systems were 6-volt and positive-ground, with the vehicle's chassis directly connected to the positive battery terminal. Today, all vehicles have a negative ground system, except for a few of our classic British cars, and maybe a few others. The changeover from 6 to 12 volts happened during the 1950s when bigger engines with higher compression ratios required more electrical power to start. Smaller cars, which required less power to start, stayed with 6-volt systems into the mid-1960s, such as the Volkswagen Beetle.

Each cell in the lead-acid battery provides 2.1 volts for a total of 12.6 volts at full charge. Each cell consists of alternate plates of lead (cathode) and lead coated with lead dioxide (anode) immersed in an electrolyte of sulfuric acid (anyone remember buying a battery at the store, and along with it came a container of sulfuric acid to pour *very carefully* into each cell?). The acid reacts with the plates, changing their surface to lead sulfate as the battery discharges. This chemical reaction releases electrons, allowing them to flow as electricity, but this reaction occurs only when there is a place for the electrons to go (such as to the starter). Otherwise, the lead would be consumed quickly. When the battery is recharged, the chemical reaction reverses and the lead sulfate reforms back into lead dioxide. With the plates restored to their original condition, the process may be repeated over and over.

The primary reason the battery wears out is because not all the lead gets returned to the plates during recharging. The lost material from the lead plates accumulates at the bottom of the cells and eventually may short-circuit the plates. Excess heat is the other cause of battery failures, as it causes the water in the electrolyte to evaporate, decreasing the surface area of the plates exposed to the electrolyte. (That gas station attendant who checked and topped-off your battery wasn't putting in acid, but distilled water.) Modern car batteries have reduced maintenance requirements, and may not even provide caps for the addition of water to the cells. Such batteries are made with extra electrolyte above the plates to allow for losses during the battery's life.

In the past, batteries required maintenance in the form of checking the electrolyte level. Modern batteries retain their fluid for the life of the battery. A weakness of these batteries is that they are very intolerant of a deep discharge, such as when the car battery is completely drained by leaving the lights on. This coats the electrodes with sulfate deposits and can reduce the battery's lifespan by a third or more. Corrosion at the battery terminals can prevent a car from starting due to electrical resistance, which can be prevented by the proper application of dielectric grease.

Here are a few things to keep in mind when purchasing a new battery:

> Batteries are grouped by physical size, type and placement of the terminals, and mounting style.
>
> Ampere-hours (A·h) is a unit related to the energy storage capacity of the battery.
>
> Cranking amperes (CA) is the amount of current a battery can provide at 32°F (0 °C).
>
> Cold cranking amperes (CCA) is the amount of current a battery can provide at 0°F (−18°C). Modern cars with computer controlled fuel-injected engines take no more than a few seconds to start and CCA figures are less important than they were in the days of carburetors.
>
> Hot cranking amperes (HCA) is the amount of current a battery can provide at 80°F (27°C) for 30 seconds and

maintain at least 1.2 volts per cell (7.2 volts for a 12-volt battery).

Reserve capacity minutes (RCM) is a battery's ability to sustain a minimum stated electrical load. It is defined as the time (in minutes) that a lead-acid battery at 80 °F (27°C) will continuously deliver 25 amperes before its voltage drops below 10.5 volts.

Manufacturing date is when the battery was produced. When batteries are stored, they can start losing their charge. A battery made in October 2015 will have a numeric code of 10-5 or an alphanumeric code of K-5. "A" is for January, "B" is for February, and so on (the letter "I" is skipped).

Despite all these great things I've just told you about lead-acid batteries, and the growing popularity of battery-powered vehicles, note that it makes little sense to use batteries to propel trains, ocean-going ships and large airplanes. Batteries are simply too heavy. Competing against fossil fuel for these applications is a challenge, and scientists are working diligently to develop compact, light-weight batteries that can store massive amounts of energy.

Here's a bit of history of lead-acid batteries, if you're interested. The lead–acid battery was invented in 1859 by French physicist Gaston Planté and is the oldest type of rechargeable battery. Prior to this development, batteries could not be recharged, meaning they were discarded after the chemicals in them were consumed (think flashlight batteries).

But, the first battery may have been about 2,500 years old! It supposedly was discovered at a Mesopotamian archaeological dig near Baghdad, Iraq in 1936. The 6-inch-tall clay pot was sealed with bitumen (a tar-like substance) and contained an iron rod surrounded by a copper cylinder. A number of similar specimens have been found. Experiments demonstrated that when filled with a liquid electrolyte such as lemon juice or vinegar, the vessel could act as an electrochemical cell, producing as much as 2 volts!

Since they didn't have cell phones back then, or even flash lights, why did they need batteries? Archaeologists disagree, but it has been speculated that the cells might have been used to electroplate metal, a process in which a thin layer of gold or other precious metal is coated onto the surface of another material. The Egyptians are said to have electroplated antimony onto copper over 4,300 years ago, and the Parthians, who ruled Baghdad (ca. 250 BCE), may have used batteries to electroplate silver.

Compiled from various internet sources.

Automotive Battery Explosion

A few years ago, my neighbor called me in a panic. He'd heard an explosion in his basement and the air was filled with smoke and strong fumes. I came running over to investigate. I recognized the smell as sulphuric acid. I asked if he had anything with a lead-acid battery and a charger, although I already suspected the problem. The battery on his back-up sump pump had exploded.

After airing out the basement, we inspected the battery. The top of the battery had blown right off, landing at the other side of the room, and exposing the lead plates, not to mention splitting open the protective case that housed the battery. I'd heard of this happening, but this was the first time I'd ever seen one.

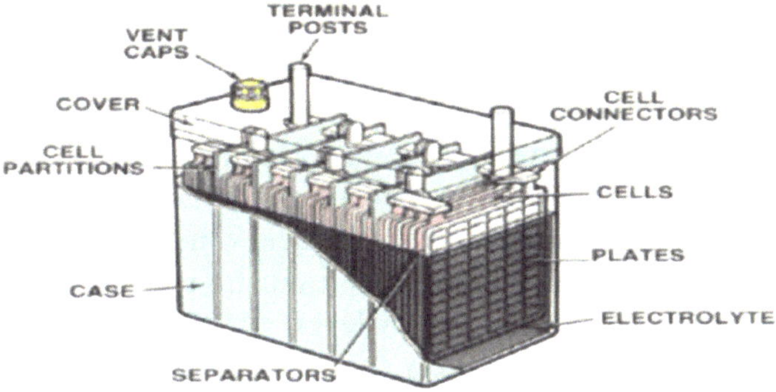

So, why did this happen and what can one do to keep it from happening again? Let's look at how a typical lead-acid battery, as used in cars, garden equipment, and backup sump pumps is constructed and works. The figure above shows a typical automotive battery. An outer polymer case acts as a container for an electrolyte (sulfuric acid) and lead plates. Each cell delivers 2.1 volts, and connected in series, the battery with six cells produces a total of 12.6 volts at full charge. The reaction between the acid and the lead causes electrons to flow and, in the process, creates lead sulfate, which stays in solution. During charging, an electrical charge is placed across the battery, reversing the process, depositing the lead back on to the plates. However, during the charging cycle, hydrogen gas is generated and accumulates in the space above the acid.

Normally the hydrogen is vented through a small hole in each cell. These gases can be ignited easily by a flame or spark. If the hydrogen is ignited inside the battery, the vent holes cannot relieve the pressure quickly enough and the top of the battery case blows off, showering sulfuric acid in the immediate vicinity along with fragments of the battery case.

So, why should there be a spark inside a battery? During the charging process, a small amount of lead doesn't get deposited back onto the lead plates, but will settle out as free lead particles. Thus, the plates slowly erode away over time, possibly loosening the connections. If the electrolyte is low, exposing the electrical connection to the air and hydrogen in the top of the case, and the connection is eroded, then the charging current can arc across the bad connection, igniting the hydrogen gas. If the electrolyte is at the proper level so that all plates are covered, any loose connection will be below the fluid level and there will be no spark.

How does one prevent either of these things from happening? If your battery has a plug over each cell, check the electrolyte level regularly and make sure it is over the top of the plates. If low, top it off with distilled water only; do not add additional sulphuric acid (let a shop check the density and add more acid if needed). As far as the connections breaking down, there isn't much you can do about that. By keeping the fluid at the proper level, all the connections should be immersed in the electrolyte, so a worn out battery should simply fail with age and not explode. This usually happens in a cold parking lot at midnight (or when trying to start your MG when you're already late for a car show) when you turn the key and hear a click-click-click, or nothing at all. There might be enough power to run the lights and horn, but not the starter.

As it turned out, my neighbor's battery was eight years old and he didn't know anything about checking the electrolyte level. Even though it'd never been used except to test the sump pump occasionally, the battery was well past the age at which it should have been replaced. Even if he'd added water as needed over the years, the plates probably were shot to the point that one or more of the cells weren't producing full power. A word to the wise, check the electrolyte level often and replace the battery after so many years; don't wait for the battery to die or explode.

Thanks for information for this article from Ed Mosher, owner of the former Kettering Automotive in Dayton, Ohio.

Picture and additional information came from: http://www.propertycasualty360.com/2013/01/23/i nvestigating-automotive-battery-explosions?t=investigative-forensic&page=3

Ammeter vs Voltmeter: Which is Better?

For years, I wondered why car manufacturers changed from an ammeter to a voltmeter. I always thought that an ammeter was better. The volt meter seemed to just sit at about twelve volts all the time, hardly moving, while the ammeter told me what current actually was flowing into or out of the battery. I thought it was good to know if the battery was either charging or discharging. As it turns out, I was wrong (it happens from time to time, and I'm always willing to fess up to it). The change came at about the same time cars switched from a generator to an alternator, and the two changes are connected.

Most people are a bit intimidated by electronics, but we can make an analogy to an oil system (or any liquid, for that matter). Two measures of importance are pressure and flow rate. Obviously, it takes some amount of pressure to get the oil to flow through a pipe. If everything else remains unchanged, it takes more pressure to get a higher flow rate. Thus, the functions of the pressure gauge and flow rate gauge to help judge the system's operation should be obvious. By comparison, volts and amps are to an electrical system what pressure and a flow rate are to an oil system. A voltmeter measures electrical "pressure" in volts and an ammeter measures electrical "flow" in amps. While I'm no authority in electronics, the real experts state that by knowing the basics of how the electrical system works, volts (i.e. pressure) gives a more useful indication of overall health of the car's electrical system than amps (i.e. flow rate). I'll go into more detail on this later.

Before I go any further, it might help to understand the basic differences between a generator and an alternator. Cars used to have DC generators, but these have been completely replaced by alternators in modern vehicles. Both produce electric power by the motion of a coil of wires moving through a magnetic field, a principle discovered in the early 1800s. Generators create direct current (DC) power and alternators create alternating current (AC), but the significant differences go far beyond that.

The basic physical difference between the two is which component spins and which is stationary. In the generator, a shaft containing wire windings spins inside a magnetic field to create the electrical current. This is reversed in an alternator; the windings are fixed and the magnet is on the spinning shaft. The alternator is more efficient because the wire winding are the biggest and heaviest part of both devices, so in the alternator the magnet, which is the lightest part, is spinning. This means the alternator can work at a higher speed and also can produce more power at lower speeds (I don't quite understand why, but this is what the experts said). Alternators tend to be more reliable than generators, largely because of the difference in how each use rings and brushes to send the current out of the unit. DC generators use split rings (i.e., two half rings), which cause the brushes to wear more quickly because they rub against the ring and hit the break between the two half-rings. An alternator uses solid rings, so the brushes receive less wear and tear. Again, there's more to the design differences than I could figure out, but the generator produces little voltage at idle and low speed and can take about ten minutes of continuous driving at moderate or higher speed to recharge the battery after one start. Because of its greater efficiency and ability to produce output even at low speed, the alternator can charge the battery in only a few minutes even at idle.

In the old days when cars used generators, ammeters typically were used to monitor the electrical system. But generators had their limit, as they couldn't hold up to prolonged battery recharging (think cranking your antique car for a few minutes on a cold day after not having driven it for a month or more). Recall that generators had a set of heavy windings that spun inside the magnetic field. These windings were soldered onto the segments of the armature where the brushes picked up the power. Generators working at high power over an extended time period constantly charging a weak battery tended to overheat, melting the rotating armature's soldered connections. The fast rotation then threw the solder from the armature, resulting in generator failure. In other words, trying to recharge a weak or dead battery could lead to a dead generator.

The generator also had an internal voltage and current limiter to help prevent overheating, as well as a cut-out relay that disconnected the generator when the engine was shut down. If the current limiter malfunctioned and allowed too high of an output, once again the generator's armature overheated and threw the solder. If the cut-out relay were to stick when shutting down the engine, it would short circuit the battery, causing it to discharge (and possibly overheat and start a fire).

In addition, generators provided little, if any output at idle and low rpm, as mentioned earlier. As a result, current flowed into and out of the battery on a regular basis. The ammeter often swung from one side to another as it measured the amount of current and in which direction it flowed. If the ammeter showed a prolonged discharge at driving speeds, that indicated a charging system problem because current was flowing out of the battery when the generator should be replacing it.

Compared to the generator, an alternator has less tendency to overheat, and having stationary windings permit it to work harder over an extended time. A large amount of air is drawn in by the alternator's fan and circulated over the windings, plus (since the heavy windings do not rotate) melted solder wouldn't be thrown about even if the alternator were to overheat. Again, I don't quite understand why, but alternators don't need a current limiter to control the amount of current output, but only need a voltage limiter. Since alternators are able to provide output at idle and low rpms, and thus better able to keep the battery fully charged, the constant flow of current into and out of the battery is greatly reduced, minimizing the information that current flow alone told the driver about the electrical system's health.

To interpret what the voltmeter is telling us, let's take a closer look at the car's electrical system. Recall that the battery is needed to start the engine, operate accessories while parked, filter any spikes in the system, and provide limited electrical power should the charging system fail. Also, recall that lead-acid automotive batteries are good at producing the high amperage needed for start, but need to be recharged before getting too depleted, or they won't recover completely. Automotive batteries contain six cells, capable of delivering 2.1 volts apiece when fully charged (This is a function of the chemical properties of lead and sulphuric acid.). Although we call it a twelve volt battery, it actually produces 12.6 volts. With the engine off, a healthy battery should read 12.6 volts. (Note that for each 0.1-volt drop, the battery loses about 20% of its ability to provide the high current needed for starting, so if it's reading only 12.0 volts, its ability to start the engine is greatly degraded.) During engine start, voltage will drop momentarily to around 11 volts which is normal because of the huge load placed on the battery. With the engine running, system voltage should be one to two volts higher than the engine-off reading, depending on the car's design.

So, here's what you need to know based on what the voltmeter tells you. In order to keep the battery charged, the generator or alternator needs to produce about 14 volts, which is what the voltmeter normally should read while driving. Close monitoring of the volt meter will show some variations, which are normal. It might drop to around 13 volts while idling with lights and other electrical devices in use, or in the case of a discharged battery being recharged. Voltage might rise to close to 15 volts at higher engine speed with minimal lights or accessories in use. (Do you recall from the old-old days that people used to turn on their headlights while driving on the freeway to keep the battery from being over-charged?) But here's what you especially need to watch for: lower than 12.6 volts while driving indicates the alternator is failing to keep up with electrical usage, or, it has failed and the battery is supplying electricity to the ignition system, which it only can do for a few hours (I can verify this from personal experience). Also, higher than 15 volts indicates a fault in the charging system's voltage regulator or related circuits. If it's substantially excessive, this can cause damage to the battery or elsewhere in the electrical system. Overcharging can generate hydrogen sulfide gas (it smells like sulfur) and cause the battery to bulge and even explode. Driving with a dead battery and relying solely on an ammeter, you'll probably see a positive reading and think everything is fine as there is current flowing into the battery, but the battery isn't taking the charge.

So, while the voltmeter may not appear to move much, those small variations actually tell us much about your electrical system's health. Usually, it is safe to assume that if the voltage is between 13.8 and 14.2 (for a 12 volt electrical system) while the car is in motion, then the battery is fully charged and the electrical system is working properly. This may be the reason most manufacturers eventually replaced the voltmeter with an idiot light that illuminates when the system voltage falls outside the range of 13.8 to 14.2 volts, since most drivers probably ignore a gauge that seldom moves.

If I rambled too much in the previous paragraphs, here's a brief summary for interpreting the voltmeter:

12.6 volts with accessory switch on but engine not running: good
Less than 12.6 volts while cranking engine: normal
Less than 12.6 volts while engine running: problem
13.8 to 14.2 volts while engine running: normal
As low as 13 volts while idling: normal
Over 15 volts while engine running: serious problem

Auto manufacturers continued to install ammeters in the dash for some years after switching to alternators because the car-buying public was accustomed to seeing an ammeter in the instrument panel. If you own an old car and want the improved monitoring capabilities and the added safety margin afforded by a voltmeter while retaining the original looks of your classic instrument panel, gauge repair and restoration shops such as Williamson's Instrument Service (www.williamsons.com) and Redline Gauge Works (www.redlinegaugeworks.com) can convert your ammeter into a voltmeter while keeping its original appearance. In the event you are adding a lot of constant high-output accessories such as electric fans, an electric fuel pump, or a high-amp stereo, you might want to consider upgrading the wiring as well. Such upgrades might be available from MAD Enterprises at www.madelectrical.com.

Based on articles by Marlan Davis at Hot Rod Magazine, Rob Siegel at hagerty.com, Mike Southern at sciencing.com, Randy Rundle at fifthavenuegarage.blogspot.com and an online post by Brad Bergholdt of Evergreen Valley College in San Jose CA.

Voltmeter Follow-Up

After writing the previous article, I decided to add a voltmeter to my TD. I will not modify the instrument panel so as to maintain originality. But, I already had a water temperature gauge that I installed on the bottom edge of the instrument panel, so as an experienced woodworker, it was a simple task to make a new mount that incorporated both gauges and matched the walnut dash panel.

After hooking it up, here are the results I saw, and a comparison to the voltage readings I expected to see

	Expected Value	Actual Value
Engine off/ignition on	12.6	12.0
Cranking	~11	10.0
Idle	~14	11.5
Fast idle	13.8-14.2	13.8
Idle with lights on	~13	10.5
Driving at moderate speed	~14	14

Note that all readings are lower than expected, except for the last one. There probably are at least three reasons for this:

1. Gauge Errors. The gauge may not be that accurate. It only cost $18. There were others listed that cost 2 – 3 times as much. They might be more accurate, maybe not.

2. Poor Electrical Contact. In my hurry to see what would happen, I didn't make any effort to clean the terminals where I connected the wires. After thirty years of running since the car was restored, bits of corrosion and dirt likely have found there way onto the terminals.

3. Location of Terminals. There are lots of places to pick up the voltage signal. I ran the wire from the gauge through the firewall and attached one wire to the cable going from the positive battery terminal to the chassis (remember, the TD has positive ground), and the other from one of the terminals on the fuse block. There likely are losses through the wiring, as all wire produces resistance. There might also be a voltage loss going through the fuse. I could have picked another spot, such as on the key switch, which is before power even gets to the fuse block. But, I didn't want to pull the switch, and it's getting too hard for me to crawl under the dash to work from behind (I'm not as young as I used to be).

While these readings were lower than expected, they all followed the trend I was expecting. For now, I'll just say that it is important to know your car's normal readings and watch for any deviations from them.

How Clutches Work

You probably don't give your clutch much thought except for when it goes bad. I'll try to explain a bit about how it works and what it does. A clutch simply is a pair of disks that rub against each other in order to take the power from one rotating shaft and transfers it to another one. Many pieces of heavy equipment and tools have clutches. When the car is stopped, the engine keeps turning but the transmission and wheels do not. The clutch is needed to transfers power from the engine and gradually bring the transmission and drive wheels up to speed. In order for a car to stop without killing the engine, the transmission then needs to be disconnected from the engine as the car slows to a stop. (We've all forgotten at least once to push in the clutch pedal as we slow to a stop, haven we?) The clutch allows us to smoothly engage the spinning engine to a non-spinning transmission by controlling the slippage between them.

To understand how a clutch does this, it helps to know a little bit about friction, which is a measure of how hard it is to slide one surface over another. Friction is caused by the minute peaks and valleys that are part of every surface – even very smooth surfaces still have this roughness, although at the microscopic level. The larger these peaks and valleys are, the harder it is for the two surfaces to slide past each other. A clutch works because of friction between the clutch plate and the flywheel. The actual amount of force generated by the friction is a function of the characteristics of the two materials rubbing against each other (called the coefficient of friction) and the force with which the two materials are pushed against each other.

In a car's drive train, a flywheel connects to the engine and a clutch plate connects to the transmission. When your foot is off the clutch pedal, springs push the clutch disc, which in turn presses against the flywheel. Friction holds the engine to the transmission shaft, causing them to spin at the same speed. The amount of power that can be transmitted from the engine to the transmission without the two surfaces slipping depends on how much force the spring puts on the clutch to hold it against the flywheel. The friction force between the clutch and flywheel works the same as a brake shoe against a brake drum or a brake pad against a rotor (for you folks with modern cars).

While driving, the spring keeps the clutch plate pressing hard against the flywheel. When the driver presses the clutch pedal, it decreases the force between the clutch plate and flywheel and allows slippage between them. Note the springs in the clutch plate. These springs help to isolate the transmission from the shock of the clutch engaging.

This design usually works pretty well, but it does have a few drawbacks. The most common problem with clutches is that the friction material on the disc wears out. The friction material on a clutch disc is very similar to the friction material on the pads of a disc brake or the shoes of a drum brake – after a while, it wears away. When most or all of the friction material is gone, the clutch will start to slip, limiting the amount of power it can transmit from the engine to the transmission.

The clutch only wears while the clutch disc and the flywheel are spinning at different speeds while engaging. When they are together firmly, the friction material is held tightly against the flywheel, and they spin at the same speed. It's only when the clutch disc is slipping against the flywheel that wearing occurs. So, if you ride the clutch a lot, you'll wear it out a lot faster.

Sometimes the problem is not with slipping, but with sticking. If your clutch won't release properly, it will continue to turn the transmission shaft. This can cause grinding, or completely prevent your car from going into gear. Some common reasons a clutch may stick are a broken or stretched clutch cable, a misadjusted linkage, or improper clutch components.

A "hard" clutch is also a common problem. All clutches require some amount of force to depress fully. If you have to press too hard on the pedal, there may be something wrong. Sticking or binding in the pedal linkage, cable, cross shaft, or pivot ball are common causes.

I'm sure there is a lot more to know about clutches.

I borrowed extensively for this article from the write-up by Karim Nice and Charles Bryant that I found at: https://auto.howstuffworks.com/clutch.htm

Evolution of Automobile Map Guidance Systems

With modern GPS-based navigation devices, most of us give little thought to finding our way to a new location (btw…GPS stands for Global Positioning System, a constellation of twenty four satellites orbiting the earth, plus a few spares in orbit, just in case you didn't know). But the first automated in-car navigation system was developed long before we had the technology to put anything into space.

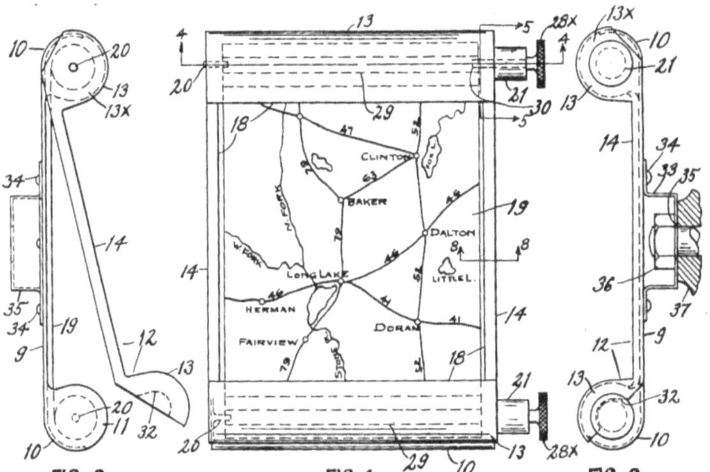

FIG. 2. FIG. 1. FIG. 3.

I'm sure most of us have used folded-up paper maps, and if you grew up in the 50s – 70s and went on a road trip, you'll remember the AAA Trip-Tick, a custom made booklet of map pages that you flipped through to guide you from point A to point B.

Paper maps could be cumbersome to use at times, requiring the driver to shift attention and hands between driving and folding a map (we've all fumbled with this, haven't we?). But the concept of the modern navigation system can be traced back to a US patent granted in 1921 to John J. Bovy. It was a dashboard-mounted navigation aid that used maps printed on rolls. As best as I can figure, the driver turned a dial manually to advance the map as needed. When you drove past the area shown on the map, you inserted another roll that picked up where the last one left off.

A few years later, in 1927, a British firm manufactured the Plus Four Wristlet Route Indicator, which was like a wrist-mounted version of the above. It came bundled with rolls covering many popular travel routes across the UK.

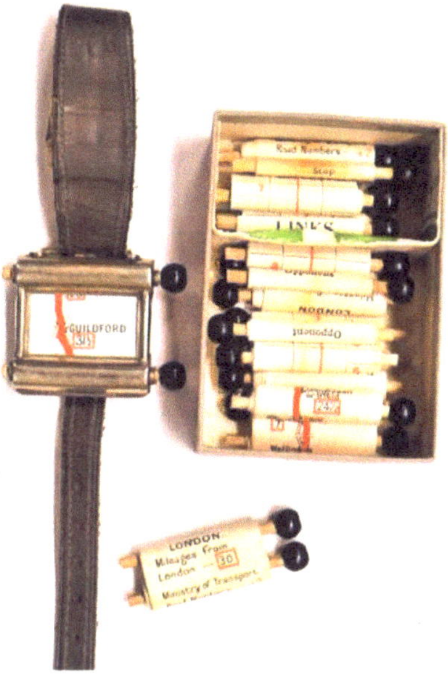

The next big step forward came in the early 1930s with the creation of the Iter-Auto, manufactured by an Italian company. The device was mounted to the car's dashboard and loaded with routes printed on long paper rolls. But the big advance was that it was hooked up to the vehicle's speedometer, so the roll would advance automatically in proportion to the car's speed and distance traveled. The maps themselves also included alerts of upcoming road features, like bridges and railroad crossings, as well as garages, hotels and such. With its connection to the speedometer, this level of automation offered a new hands-free approach to (very) roughly accurate navigation. Since the device's only input was speed, and not direction, and maps printed were only useful for routes along predominantly straight roads. I don't know how accurately the device maintained the current location at the center of the display, but I'd assume there was a manual adjustment to reposition the roll as it got off center. Turning onto an intersecting road meant loading another roll with the new map. It's not surprising that the Iter-Auto never achieved mass-market success. Thus, it's not surprising that paper maps remained the standard for navigation for the next fifty or so years.

That said, the idea of combining telemetry with physical maps re-emerged in 1981 with the Honda Electro Gyrocator, but it proved to be a prohibitively expensive optional extra on only a few Honda models of the time. A driver would insert a transparent map over the 6-inch monochrome screen and move a dot to the starting location on the map. The dot then moved around the map overlay based on information fed to the device by the car's transmission and a motion sensing device contained within. Makes me think of the very first generation of home video games in which the computer generated only a moving dot and the user placed a sketch of a soccer field, ping pong table, or whatever on the monitor.

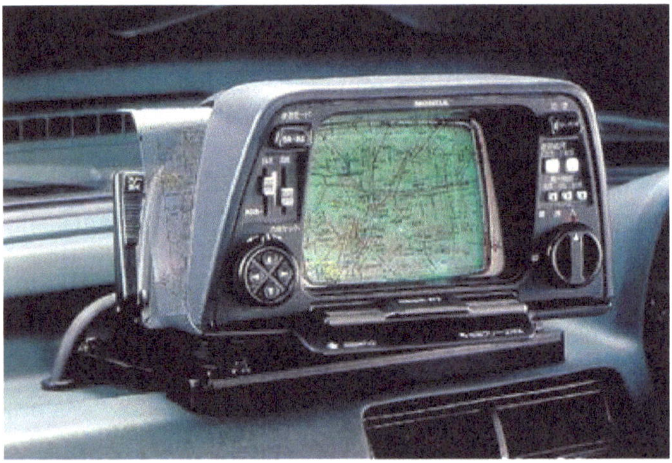

Unlike most navigation systems of today, the Honda Electro Gyrocator did not use GPS satellites to determine its position and detect movement of the vehicle. Rather, it used a simplistic, but ingenious inertial navigation system. These systems already were in use on airliners and were very accurate, but extremely expensive.

They used a spinning gyroscope that could sense any change in speed or direction, left, right, or vertical. But these precisely-made gyros were way too expensive for the automotive industry. Honda's solution was to direct a stream of helium at two heated wires within a sealed capsule. When the car turned, inertia diverted the stream, hitting one wire more than the other. The temperature difference between the two wires was interpreted as a direction change. (Point of interest – Try this…point your car's vent at your face, then make a sharp turn. You'll feel the stream of air move to one side.)

The Electro Gyrocator couldn't choose a route for you, but it could trace a line on the screen, and in 1981, that was a big step forward. Users still had to place the appropriate transparent map overlay in front of the screen, swapping it out as needed. When it went on sale in Japan on Accords and Vigors in 1981, the buyer received a book of these maps covering all of Japan (thus, I assume the system was available only in Japan). But, the device was not a success, as the Electro Gyrocator's capabilities were too limited and its price too high. At $2746 USD, it was almost a quarter of the car's cost. It is not clear how many units were sold. But it formed a basis for future GPS-based systems that would arrive soon after the Electro Gyrocator's demise.

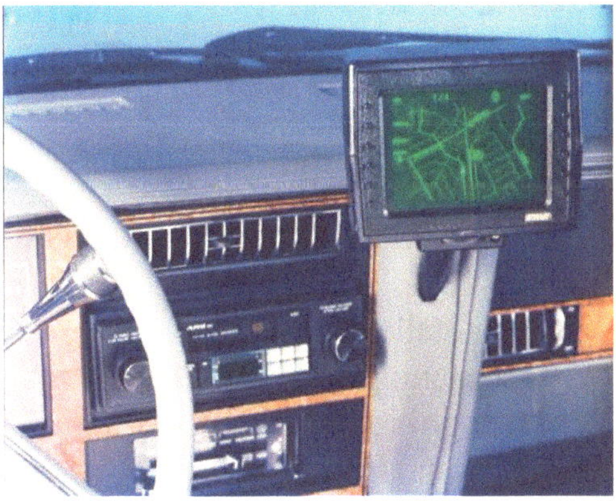

The Honda Electro Gyrocator was followed a few years later by the Etak Navigator. This system used inputs from wheel sensors and a compass to determine location, and displayed an actual digitalized map. The maps were input using cassette tapes, and required six such cassettes to cover the San Francisco area, the first map offered, but soon was followed by maps for other major metropolitan regions.

Because the Etak Navigator could turn street addresses into longitude/latitude coordinates, it could actively guide the driver to the destination. The Etak Navigator first sold around July 1985, with the 450 model retailing for $1,395 (about $3,083 when adjusted to 2015 dollars) and the 700 selling for $1,595 (the difference was the size of the display). Map cassettes cost about $35 each (not sure if this was per cassette, or per metropolitan area). Local car stereo dealers installed and calibrated the system. Etak changed hands and names several times over the years, and eventually was acquired by TomTom, one of the early makers of GPS navigation systems, namely for access to the extensive set of digital maps Etak had created.

However, even before the Etak Navigator was released, numerous companies were conceptualizing the first satellite-based car navigation systems. When the full GPS system was made public in 1995, all prior systems became obsolete rapidly. (GPS was developed for military use and was accurate to within a few yards. Initially, a downgraded signal was available for civilian use, but still was accurate to abut 100 yards). Indeed, even standalone navigation devices faded as auto makers offered navigation systems integrated into the car's dashboard electronics. Or, if your car is more than a few years old, just call up an app on your smartphone. What's next? Who knows!!

Sources:

1. https://www.motor1.com/news/137950/honda-electro-gyrocator-ieee-milestone/
2. https://www.roadandtrack.com/car-culture/a36355103/the-first-automotive-gps/
3. https://imgur.com/gallery/QOUtH2z
4. https://www.geocaching.com/blog/2014/11/before-gps-and-geocaching-existed-three-navigation-systems/
5. https://www.fastcompany.com/3047828/who-needs-gps-the-forgotten-story-of-etaks-amazing-1985-car-navigation-system

Automotive Safety Glass

Glass has been around for thousands of years. It probably was discovered when someone lit an especially hot fire on the beach, causing the silica in the sand to melt and congeal into a shinny blob as it cooled. People realized that they could use these shinny blobs for useful items, such as beads for jewelry, eventually learning how to make useful items such as plates and bowls. They also discovered that by adding certain other elements into the molten blob, either intentionally or unintentionally, that they could make the blob translucent, and eventually, even just about clear. By the middle ages, glass was used for windows, but it wasn't very clear nor was it smooth. However, it was the perfect medium for stained glass windows, as different chemicals mixed with the molten glass produced vibrant colors as the sunlight shown through.

Glass is not a solid, but neither is it a liquid—supercooled or otherwise. It is called an amorphous solid—a state somewhere between those two states of matter.

Jumping ahead a few hundred years, the process was refined enough to make clear glass windows. Early cars were little more than motorized carriages, but drivers soon demanded more protection from road hazards like mud and flying rocks, especially as speeds increased. Windshields were introduced in 1904; most were a horizontally-divided piece of plate glass, the glass being just like the glass used for house windows. When the top half got too dirty to see through, a driver could fold it down. By 1913, Reo offered a windshield as optional equipment. In 1915, Oldsmobile was first brand to include the windshield as standard equipment.

As cars became more popular and affordable, a rise in accidents was inevitable. When one of these early cars was involved in an accident, it was not uncommon for the occupants at a minimum to be injured by flying shards of glass, or far worse, lose their life after going headfirst through the windshield. There are stories that even Henry Ford was injured by flying glass in an accident. Sadly, seat belts, which would have at least kept occupants in their seat, still were a half-century away.

There are two great advances in glass-making that greatly improved the safety of automotive glass. The first was tempered glass. Tempered glass is made by cooling the outsides of the sheet of glass quickly while the inside cooled more slowly. This results in different amounts of stress within the finished sheet of glass, so that when the sheet breaks, it splinters into tiny fragments that are much less dangerous than large chunks. As with many great discoveries, this one also came by chance. A glass-maker allowed a blob of molten glass to drop into a vat of water, quickly cooling the outside layer. He (or maybe it was a she, I don't know) noticed that he could hit the blob with a hammer and it wouldn't break, but if he hit the tail end that hadn't been in the water and had cooled normally, the entire form shattered into uncounted tiny pieces. I have no idea why he hit it, but fortunately he realized his discovery. (Note – I once disassembled an oven door to clean all the nooks and crannies. The glass slipped out and hit the floor, and the next thing I saw was a pile of glass pieces resembling a mound of rock salt.)

The second innovation was laminated glass. Two European inventors developed glass laminating, Frenchman Edouard Benedictus and Briton John C. Wood. Benedictus made an accidental discovery in his lab. As the story goes, in 1903 he was working with cellulose nitrate, the clear plastic used for movie film, and put a flask he'd been using on a shelf. Benedictus bumped that shelf, knocking the flask to the floor. It broke — but the pieces held together, pretty-much retaining its original shape. He eventually concluded that the cellulose nitrate left in the flask had dried and held the glass fragment together. After experimenting further, Benedictus developed safety glass – two layers of plate glass with a layer of cellulose between them, and promoted its use in automobiles. Unfortunately, it was costly and both manufacturers and drivers had a strong interest in keeping cars affordable. Though Benedictus was granted a patent in 1909, the product was not put into use until World War I when laminated glass was used in the goggles of gas masks. Meanwhile John Wood in England also had been working with cellulose and devised another method for adding a protective layer (originally tree resin, later cellulose) between two pieces of glass and also creating shatter-resistant glass. His method was patented in 1905, ahead of Benedictus. In late 1919 Ford began using laminated glass, and directing its use in all Ford cars over the next decade. It eventually became the standard for all manufacturers around the world.

Another aspect also helped improve safety. A urethane glue was used to bond the glass to the frame. When an unfortunate occupant was thrown against the laminated glass, the shattered pieces remained in place, keeping passengers from being ejected. Its strength even added a bit of structural integrity for the car if it rolled over.

In 1926 Rickenbacker offered safety glass as a standard feature all the way around its car. By 1937, safety glass became mandated for all cars. Changes occurred slowly throughout the years with innovations like the curved windshield from Studebaker in the 1940s, making the manufacture of laminated glass more difficult. Manufacturers began using tempered glass for side and rear windows in the 1950s.

One significant problem with early laminated glass, however, was that its inner celluloid layer would discolor, darken, and become brittle over time. In 1938 Carleton Ellis patented a clear synthetic resin that did not discolor over time. Beginning in the late thirties, manufacturers started using Polyvinyl Butyral (PVB) which made laminated glass clearer and stronger and even helped block high frequency sound and harmful ultraviolet rays.

Beginning in 1966 cars came equipped with improved laminated windshields that could withstand nearly three times the impact of earlier versions. In the 60s and 70s, Federal Motor Vehicle Safety Standards were set for the strength and clarity of laminated windshields (FMVSS 205); windshield retention strength during accidents (FMVSS 212); roof rigidity in rollover accidents (FMVSS 216); and limits on windshield penetration (FMVSS 219).

Nowadays windshields continue to grow in complexity and sophistication. Windshields are larger and often more raked. Some provide increased visibility with a windshield that extends up into the roof above the driver or one that wraps around to the side of the car. Modern windshields can filter 95-99% of ultraviolet rays. Since the 90s, a hybrid film with dye to absorb heat and microscopic metal particles to reflect sun rays has provided significant reduction in infrared rays and consequent internal heat gain.

Today, laminated glass is required for all front windshields, but, despite continued government pressure to change, tempered glass typically is used for side and rear windows. There's a reason for not using laminated glass all around. What if the car should roll over, jamming the doors shut, or the car should wind up in deep water and the occupants can't force the doors open? Using a sharp object, the occupant can shatter the side glass and escape. It also would be easier for rescuers to cut into a vehicle to free trapped passengers. That's what the manufacturers claim, anyway.

Sources:

1. http://www.tpglassworks.com/auto-glass-services/a-look-at-the-invention-of-automotive-safety-glass/#:~:text=Automotive%20safety%20glass%20can%20trace,was%20using%20for%20his%20experiment.
2. https://www.windshieldexperts.com/blog/101/
3. https://www.libertyautoglass.us/auto-glass/windshield-glass-repair-effort-pa-history-auto-safety-glass/
4. http://www.secondchancegarage.com/public/windshield-history.cfm
5. https://www.hg.org/legal-articles/the-sordid-history-of-auto-safety

Road Debris Alert

With the development of modern tires, flats, especially blowouts, have become much less frequent than in previous decades. A punctured tire can take weeks until it goes completely flat, and a soft tire often is noticeable by a reasonably-aware driver before it becomes flat.

While having a tire puncture repaired, my mechanic mentioned that he noticed a significant increase in the number of tire punctures in the last year or so. He credits this to a hail-storm that hit our area two years earlier. That storm damaged many roofs, resulting in roofs being replaced at a significantly greater rate than normal. In my block alone, over half the roofs were replaced that summer. Lots of debris, including nails, wound up in the streets, eventually finding its way to the curb.

As a good driver, you're careful and keep plenty of clearance from the curb, right? Guess again. He told me that a disproportionate number of the damaged tires he saw were on the right rear. You'd expect the front tire to hit the nail first, right? But, think about the turns you make. Left turns usually are from an inside lane to an inside lane. You're relatively farther away from the curb and debris such as nails. Now, let's look at a right turn. You're probably careful not to over-steer, keeping the front tire away from the curb. But, even a careful driver feels the right rear scrape the curb from time to time. That's because the rear tire tends to drift further to the right than the front tire. (Next winter, when you're the first one to drive down your street after a snowfall, make a right turn, then go look at your tire tracks.) You're probably not aware of how many times the right rear barely misses the curb because you never hear or feel the tire scraping!

The solution? Give yourself a little extra distance from the curb when starting a right turn, then make your turn a little wider than normal to keep your rear tire well-away from the curb. In addition, he advised that those of us using 1950's-style bias ply tires for authenticity should pay special heed, since these tires aren't nearly as nail-tolerant as modern tires.

LIQUIDS

A Brief History of Lubricants

You probably don't give your engine oil much thought, except when it is time to change it or when the oil light on the dash illuminates. Today's oils do much more than just lubricate moving parts to minimize wear...they provide corrosion protection, carry heat away from the engine, and much more.

The use of some sort of greasy material to lubricate moving parts certainly goes back many thousands of years. According to archaeological discoveries, it dates back to at least the seventeenth century BCE in ancient Egypt, when olive oil was used to help move large stones. By the fourteenth century BCE they used animal fats (tallow) to lubricate chariot axels. In places around the world where oil seeped out of the ground naturally, I'd assume people used that, and even developed some crude refining techniques to extract the heavier grease from the oil. For the next few thousand years, since nothing moved faster than a horse could run, there was little need for a better lubricant. However, the use of animal and vegetable oil based lubricants diminished after the first oil well was successfully drilled in Titusville, Pennsylvania, in 1859.

The demands on the lubricants started to change with the invention of the steam engine and its need to lubricate sliding and rotating metal parts at higher temperatures and pressures. Dr. John Ellis is credited as being the first person to devise a way to reduce friction in large steam engines using a petroleum lubricant.

Motor oils today solve the same problems they did back in the 1800s, but with a far better standard and consistency. They also provide lubrication at much higher temperatures and pressures than was needed with steam engines. With the invention of the internal combustion engines, the first lubricants were pitched by Dr. Ellis' company, Valvoline. Henry Ford was an early customer. Ford set a land speed record in 1904, driving a primitive machine called the "999". His up-to-100 horsepower engine gave the 999 a top speed of 91.37 mph on a frozen lake in Michigan. A few years after his speed record, Henry Ford introduced the Model T with a specific warning on its dashboard:

"This car is filled with Valvoline light motor oil. We recommend its use. No other oil should be used in this car."

During the 1920s, in response to growing demands, particularly from the fast-growing auto industry, lubricant manufacturers began processing their petroleum-based oils to improve lubricant performance. Various treatment processes, some more successful than others, were developed during that decade. Additives were developed to inhibit oxidation, resist corrosion, enhance pour points (the temperature below which the oil is too stiff to flow properly), improve viscosity indexes, and more emerged in the 1930s.

But, one of the first big developments was to create a near-universal motor oil that could be used in any engine. Early manufacturers included oil recommendations for vehicles. By the 1930s, there were hundreds of active and defunct car manufacturers in the U.S. alone, all supported by a relatively spotty network of repair facilities. Imagine trying to find the correct motor oil for a Bugatti in the middle of Nebraska! Then, Valvoline released the nearly-universal X-18 motor oil in 1939, and it solved a problem drivers had been facing for more than 30 years, namely, an oil that was suitable for a huge range of vehicles. But, many challenges still lay ahead. Different uses dictated different lubrication needs; cruising leisurely down a country road is different than driving in a 100-mile race on a hot day.

As automobiles became more and more common in the 1930s, car makers realized the need for standardization among their vehicles. The Society of Automobile Engineers took on the task of setting standards for engine oil, classifying oils by their viscosity, or the speed at which the liquid pours. They tested various oils at 100°C (chosen because that was the temperature that oil often reached in a highly-stressed engine), and SAE 30 eventually was recognized as the standard engine oil. (Viscosity is measured by the simple yet remarkably accurate method of pouring a known volume of oil through a known size hole and measuring how long it takes to come out.)

The Second World War contributed greatly in the further

development of engine oils, driven by the incredibly-high demands of aircraft and tank engines, as even the SAE 50 failed when the engine heated up. Adding to the problem, on a cold morning in the 20 degree range, 50-weight oil was so stiff that it was difficult for the engine to crank. Some relief came as a result of plastic companies researching petrochemicals. They discovered chemicals that had the ability to produce hydrocarbon polymers, and that these polymers could prevent oil from thinning in the heat while not compromising viscosity at lower temperatures.

To briefly explain this as best I can, basic physics says that any molecule will move around faster as it heats up. That's why a warm liquid flows quicker than when cold (Think pancake syrup that was stored in the refrigerator – it is thick at first, but then flows quicker from the bottle as you get to your third or fourth helping of pancakes. That's because the bottle absorbed heat from the air in the room). The same thing happens with oil. What the researchers discovered was that at higher temperatures, their polymers produced long chains of molecules that wound around each other at higher temperatures. While the laws of physics dictated that the long molecules still had to move faster due to heat, their tangled mass prevented them from flowing faster. The end result…they acted as if they retained their low-temperature viscosity at higher temperatures!

This led the way for the introduction of the 'w' rating that is common today, as in 10W30. This means that the oil is a 10-weight at winter temperatures when the engine is cold, but then acts like a 30-weight as the engine heats up to its operating temperature.

Multi-grade oils continued to improve over the years as researchers figured out how to keep the viscosity modifiers from breaking down in high-load conditions, thinning out if it mixed with fuel residues, and age. They also developed synthetic detergents for oil, and very efficient anti-wear and anti-scuff additives.

Synthetic oils were the next big development in engine oils. While they were first used during the Second World War, they weren't that great and were expensive.

After the Second World War, high speed motorways that

allowed cars to cruise at sustained high speeds for many hours, limited only by the need to refuel or a bathroom break, blossomed around the world. But lubricants were still being challenged; metallurgy and manufacturing technology advances allowed much tighter machining tolerances between moving parts and engines produced greater power. Light weight oils were needed to lubricate the tight-fitting parts, but didn't work well with other demands of modern engines. Lighter oils, under more stress, broke down more quickly and had to be changed more often, but still were needed to allow the engine to be more responsive and fuel efficient. Car manufacturers, on the other hand, want to cut down service intervals to make ownership as cheap and easy as possible.

Help came from the aircraft industry. Synthetics worked better with gas turbine engines because of the immense pressures and temperatures at which they operated. These early synthetics were expensive, because of limited production, so were used mostly for aircraft engines. But, their introduction into the automobile market soon would change that.

Originally developed as an exercise to obtain useable fuel from oil shale and coal slack, Mobil produced a family of plastics known as Poly Alfa Olefins, or PAO. These are the basis for the majority of synthetic oils today. Their first use was in steam turbines, but, in possibly the first ever use of synthetic oil in automobile engines, they were used during the Second World War in tank engines. In the African desert, Sherman tanks that used standard Chrysler car engines and petroleum-based lubricants were frying their engine oil and seizing engines. These special lubricants proved to be very successful.

But…these early synthetic oils still had a major problem that nearly caused their demise. They caused rubber seals in the engine to deteriorate, giving early synthetics a very bad reputation. It took many years of additional research to solve this problem.

Enter the Poly-Ol Ester based lubricants. They have the advantage that your engine probably will never wear them out. Another useful property of the oil is that it does not break down in storage, as does a mineral oil. A vehicle may be left for years with the oil in the sump, and started up when needed. Added to this is the extreme stickiness of the oil, which coats all parts with which it comes into contact and does not creep off, as do other 'synthetics'. Red Line® Synthetic Oil Corporation is one producer of Poly-ol ester lubricants. Shell's line of full synthetic oils are made from natural gas, with no crude oil used at all.

I'm sure there's more to the story, but I'll stop here.

Thanks to the following sources:

1. https://www.mccluskeychevrolet.com/blog/look-interesting-history-motor-oil/
2. https://www.volvoclub.org.uk/engine_oil_history.shtml
3. https://www.uti.edu/blog/industry-relationships/everything-about-motor-oil
4. https://petrolicious.com/articles/this-is-the-hidden-history-of-motor-oil
5. https://iselinc.com/brief-history-lubrication

A Brief History of Petrol

I guess I always assumed that petrol, or gasoline, as it is known in the United States, just magically appeared with the development of the internal combustion engine. Yes, sort of, but a lot of things happened before that.

We all know (I assume) that petrol is refined from crude oil. But how did that come about? How did we know to pump crude oil and make it into the variety of products available today, especially gasoline? It all had to do with the quest for a fuel for lamps that burned cleanly and gave off a bright, consistent flame. Initially, crude oil was not considered a source for such a fuel. Typical early fuels were oils derived from animal fat and olive oil.

The earliest sources of crude oil came from oil found seeping from the ground, often called "rock oil." Someone probably heated the oil, for who knows what reason, but doing this drove off the lighter elements it contained, and left behind a thick, black goo that became hard as it cooled. Descriptions of this process were given in Islamic handbooks such as those of Muhammad ibn Zakarīya Rāzi (854–925 A.D.). The streets of Baghdad were said to have been paved with tar that was derived from crude oil obtained from natural fields in the region.

In the 13th century, Marco Polo described the output of oil from what is now Azerbaijan as hundreds of shiploads. Eventaully, Arab and Persian chemists realized that the gasses that evaporated from the heated oil were flamable and devised a way to capture and condense them back into a liquid to use as fuel. Thus, a crude refining process was developed. This process in various fashions apparently became well-known throughout Babylon, Egypt, China, and the Philippines.

Jumping ahead to the middle ages, most people still lived in the dark after sundown. Candles, with their flickering and inconsistent light, were the only illumination other than the fireplace. By the early 1800s, numerous liquid fuels had been developed, including rapeseed oil, lard, and whale oil rendered from whale blubber (and the more expensive spermaceti from the head of sperm whales), and various lamp designs were developed to burn them. By 1850, the U.S. Patent Office recorded almost 250 different patents for all manner of lamps, wicks, burners, and fuels to meet the growing consumer demand for illumination.

Before the invention of kerosene, two-wicked "burning fluid" lamps, as they were called, were popular but dangerous sources of light. Up until the mid 1800s, the most popular lamp fuel by far was the "burning fluid" called camphene, an aromatic but dangerous mixture of turpentine, alcohol, and camphor oil extracted from the wood of camphor trees. It was inexpensive and burned bright, but was volatile and the lamps could explode.

Kerosene was first produced from coal in 1846, and proved to be a better fuel for lamps. But the modern history of the petroleum industry probably began in 1854 when Ignacy Łukasiewicz began producing kerosene from oil in Krosno, Poland. However, since crude oil was available only in very small quantities from natural seepage, such limited availability restricted the uses for petroleum to medicinal and specialty purposes.

The refining of crude petroleum in North America began with drilling of the first oil wells in Ontario, Canada, in 1858 and in Pennsylvania in 1859. With the discovery of "rock oil" in northwestern Pennsylvania, crude oil became available in sufficient quantity to enable larger-scale processing systems. The earliest refineries employed simple distillation units, or "stills," to separate the various components of petroleum by heating the crude oil mixture in a vessel and condensing the resultant vapors into liquid fractions. Crude oil is a mixture of many different chemicals, each with a different boiling point, so they could be separated by heating the crude oil to the specific temperature that caused the desired component to boil, then cooling the vapor to condense it into a liquid. The primary product of this process was kerosene, which proved to be a more abundant, cleaner-burning lamp oil of more consistent quality than whale oil or animal fat. The product, or fraction, with the lowest boiling point from the still was naphtha, a forerunner of gasoline. Its initial commercial application was primarily as a solvent. Fractions with higher boiling points were used as lubricants and fuel oils.

Jumping ahead to the late 1800s and the invention of the internal combustion engine, they obviously needed a liquid fuel that could be vaporized in order to burn. I'd imagine that just about every flammable liquid that was available was tried, and naphtha proved best, but still not really adequate. Gasoline was one of the waste products of refining and had no real value (it burned too easily for use in lamps), but proved to be the better fuel for internal combustion engines.

All of the products of oil refining consist of molecules made of carbon and hydrogen, but of different combinations of these atoms. Thus they are collectively called "hydrocarbons." Early refining processes produced mostly kerosene and little gasoline, but better refining processes were developed that increased the proportion of gasoline by getting the carbon and hydrogen atoms to separate and rearrange into the desired molecules. These techniques included heating the crude to higher temperatures and under pressure, adding various chemicals to the crude oil, and using catalysts to help the process. Over the years, various additives were mixed in to the gasoline to make it burn cleaner and reduce pre-ignition.

So, why do we call it gas when we know that it is a liquid, and why does the rest of the world call it petrol? Recall that crude oil often was called rock oil. The Latin words for rock and oil are petra and oleum. The word petroleum can be traced to 14th century France, and somewhere became shortened simply to "petrol." And why do we Americans call it gasoline? I'm sure there's more than one story, but I found this one in several sources. In England in the mid 1800s, a publisher named John Cassell, but who also was an importer of miscellaneous items, imported a fuel oil that was basically gasoline, and marketed it as Cazeline (in a lot of organic chemistry, -ene or -ine is used to denote a specific type of chemical bond between atoms, such as in benzene). When others tried to sell similar products, assuming the term Cazeline was a generic name, Cassell protested, and they started calling it Gazeline, which eventually became 'gasoline." The name stuck here in America, but not in most of the rest of the world.

Sources:

1. https://www.britannica.com/technology/petroleum-refining
2. https://en.wikipedia.org/wiki/Petroleum_refining_processes
3. https://www.eia.gov/energyexplained/gasoline/history-of-gasoline.php
4. https://www.thoughtco.com/history-of-gasoline-1991845
5. https://www.adrianflux.co.uk/uk-us-car-part-names/petrol-tank-vs-gas-tank/#:
6. https://mechanics.stackexchange.com/questions/44520/why-dont-cars-use-kerosene-and-engines-optimized-for-kerosene

The History of Antifreeze

Antifreeze has three important functions in the car's engine: it keeps the coolant from freezing, it carries away excess heat to lessen the chance of the engine overheating, and it keeps the metal in the engine from corroding.

Early water-cooled engines used straight water as a coolant. Water has excellent heat transfer characteristics that made it well-suited for use as an engine coolant, but it had two limitations. First, cars often operated in locations where temperatures dropped below freezing. Water has an interesting characteristic that as it freezes, it actually expands a bit (did you ever forget that you put a soft drink can in the freezer? or, did you ever wonder why ice cubes float?). This expansion is powerful enough that it can crack the engine block. Second, the engine's heat could cause the water to boil on a hot day or under a greater power demand. In early unpressurized cooling systems, this allowed the water to boil away quickly. Boiling water also could create pockets of steam within the engine that reduced cooling even more since heat doesn't transfer into a gas as well as into a liquid.

Pressurizing the cooling system with a spring-loaded radiator cap increases the temperature at which water boils. But, for cold-weather operation, water has to be mixed with something else to lower the temperature at which it freezes.

The earliest example of automotive antifreeze is methyl alcohol which is an industrial chemical used as a solvent and for other purposes (as opposed to the ethyl alcohol that people consume). Alcohol has a lower freezing temperature than water, which made it an effective antifreeze when mixed with water. However, alcohol has a number of qualities that make it poorly suited for use in automotive cooling systems. The biggest issue is that it can accelerate the corrosion of metal in the engine. Also, given that early cooling systems were unpressurized, the alcohol could evaporate over time, requiring that it be replenished periodically, but especially before temperatures fell below freezing.

The next step in the history of antifreeze was actually taken many years earlier. A substance called ethylene glycol was first synthesized in 1856 by French chemist Charles Adolphe Wurtz. However, very little was done with the substance until the 20th century, when it was used as a substitute for glycerol in the manufacture of dynamite. Someone eventually discovered that ethylene glycol could make a good antifreeze because it mixed readily with water and that it has both a lower freezing point and a higher boiling point than water. Adding ethylene glycol in equal parts with water lowers the freezing point to minus 34° F while also raising the boiling temperature to 265° F (with a 14 PSI cap). Why not use straight ethylene glycol? By itself, it carries heat 15 to 20 percent less efficiently than pure water and although it boils at 386 degrees F, it freezes at 10 degrees F, which is not low enough to provide adequate freezing protection in cold climates. The 50-50 mix has proven to be about ideal.

Ethylene glycol was first used as an automotive antifreeze in 1926, and World War II brought it into widespread use. After the war, it became the dominant chemical antifreeze for the remainder of the 20th century.

Glycerol, another chemical, was used briefly as an automotive antifreeze. This chemical works similarly to ethylene glycol, but it was more expensive to produce. However, glycerol yet may be a good alternative to ethylene glycol in the future, primarily due to the fact that glycerol isn't toxic. Advances in production techniques may result in a more affordable product some day.

Various types of corrosion inhibitors are added to antifreeze to prevent corrosion inside the cooling system. Corrosion occurs when oxygen and dissolved minerals or salts in the coolant react with metal surfaces. Corrosion also can occur as a result of electrolysis, which is an electro-chemical reaction between dissimilar metals (such as aluminum and cast iron), or from stray electrical currents that pass though the coolant due to poor grounding between the engine and chassis. In general, corrosion inhibitors keep the coolant slightly alkaline. This keeps corrosion in check as long as there is enough inhibitor to neutralize any acids that develop as the antifreeze slowly breaks down. But over time, the inhibitors themselves break down. At this point, the coolant should be changed since corrosion can start.

Up until the mid-1990s, essentially there was one type of coolant in general use: the traditional "green" formula coolant. This type of coolant contains fast-acting but relatively short-lived corrosion inhibitors that last about two to three years, or 24,000 to 36,000 miles. Green formula coolant is still available for older vehicles, but is not recommended for newer vehicles, as I'll explain.

Traditional green antifreeze is known as IAT antifreeze, which refers to Inorganic Acid Technology. It was fortified with silicates and phosphates to inhibit corrosion by coating the cooling system's metallic parts. This was especially important when the primary metal being protected from rust was cast iron in the engine block, heads, and other engine components.

During the 1980's, as more and more vehicles began to use aluminum parts and rubber seals in the cooling system, problems began to occur. For example, silicates caused massive amounts of scale to build up when in contact with aluminum, degrading the cooling system's ability to do its job. Phosphates dried and corroded anything made of rubber, such as water pump seals. The corrosion protection of IAT antifreeze is depleted at around 36,000 miles (I assume that mileage figures are based on average driving, not the miles put on a typical classic or show car), causing corrosion once the coolant breaks down. The coolant even can start to conduct electricity through electrolysis and begin eating the engine from the inside.

Newer cars use various formulations that are different colors. Long-life coolants use various Organic Acid Technology (OAT) additives to achieve a service life of up to 5 years or longer, or 150,000 miles. There also is a Hybrid Organic Acid Technology (HOAT) formula that is dyed yellow and contains silicates for added aluminum protection (Don't ask me how the chemical wizards worked that out, as it contradicts what I stated earlier). There is no formal standard for colors, so it is best to use the formulation recommended or required by your car's manufacturer, usually found in the owner's manual, then carefully read the label on the antifreeze container. Ask the store for help if you're not sure. If the salesperson isn't sure, give a polite 'thank you' and try somewhere else. Coolants are generally considered to be compatible, however, mixing coolants of two different qualities results in a mixture of unknown intermediate quality. Some studies show that mixing IAT and OAT antifreeze actually caused more corrosion than water alone. In an emergency, it's best to use water, then flush the system and replace with the proper antifreeze as soon as possible

All antifreeze breaks down over time and must be changed and flushed at some point. If you are unsure whether it's been flushed, it probably hasn't, so it's a good idea to change. All types of coolant are available in ready-to-use premixed containers (which contain half antifreeze and half water), or in full strength containers (which must be diluted with ***distilled*** water – don't use tap water, as it is filled with silicates, minerals, and other corrosion-causing salts).

One last note…to pet lovers. Unfortunately, antifreeze's main ingredient, ethylene glycol, has a taste some animals find appealing. And it doesn't take much to be fatal for dogs and cats. As little as one tablespoon can kill a dog, while one teaspoon is sufficient for cats. At the least, it can cause vomiting, seizures, and kidney failure, so be sure to clean completely any antifreeze that spills on the garage floor or driveway.

Thanks to the following sources for all the information they provided:

1. https://www.machinerylubrication.com/Read/841/coolant-fundamentals
2. https://www.scfuels.com/the-sc-fuels-guide-to-coolants/
3. https://www.autozone.com/diy/trustworthy-advice/understanding-different-types-of-antifreeze
4. https://www.knowyourparts.com/technical-resources/engine/a-history-of-automotive-coolants/
5. http://www.crankshift.com/history-of-antifreeze/
6. https://sensibledriver.com/article/the-versatile-fluid-that-changed-the-world

Automotive Finishes

Early automotive finishes were no more than what had been used for decades on horse-drawn carriages. Not surprising, since early autos were nothing more than carriages powered by either a gas engine or an electric motor. Prior to 1924, vehicles were painted using the same carriage-builder method used for decades. An explanation of the paint process from an early advertisement of the Studebaker Brothers Carriage Manufacturing Company in South Bend, Ind. proudly stated:

"Studebaker Finish! The most perfect, lasting surface of paint and varnish possible to put upon carriages. Twenty-two successive stages of finish … appear on every Studebaker carriage. But, there are other steps typical of Studebaker thoroughness … the chemists tests of oils and pigments; the careful preparation of the wood to take its finish properly; the repeated hand rubbing (sanding and polishing) to get smoothness; the perfect drying of the coats, an item that means fifty-two days of valuable time. More important than all of these, is the long, unequaled Studebaker experience, coupled with the thoroughgoing principles practiced throughout this 101 acre plant — Studebaker Finish Endures."

It took fifty-two days to paint a carriage, and this was typical of other manufacturers. *And they bragged about it to show their level of care!!* The drying times were extremely slow, and vast dust-free drying rooms were needed. But durability was poor. The varnish topcoats often began to fail in two to three years due mostly to ultraviolet rays in the sunshine.

As early automobiles began to use sheet metal, the need for corrosion-resistant primers and new sealers arose. These new products were added to the centuries-old system already in practice. However, the time-consuming and labor-intensive methods were incompatible with assembly-line manufacturing and mass production. Clearly, something needed to change.

Keep in mind that paint – any paint – is made from various color pigments suspended in some sort of material, called the binder, to hold the pigment particles in place and attached to the surface. Existing paints available at the beginning of the 20th century were based on natural linseed oil resin as the binder. Oil resins don't dry through evaporation, but cure through a chemical reaction called oxidative cross linking, meaning that chemicals in the binder hardened through a chemical reaction instead of the evaporation of a solvent. Don't worry about the chemistry...the bottom line is that the paint took a long time to dry. Recall that Ford painted all its early cars black, but it wasn't because Henry especially loved the color. It's because black paint dries faster than other colors because it absorbs more heat from sunlight than do lighter colors, speeding up the chemical reaction. But Ford's black paint still was applied by hand to the Model T in multiple coats and sanded between coats, a process that still took about a week to complete. This caused production bottlenecks on the assembly line, and Model Ts in various stages of painting filled the factory floors.

This process bottleneck led to the first paint specifically developed as an automotive coating – DuPont Company's "*Duco*" paint. This new coating technology made a great improvement in productivity by reducing the drying time for each coat from many days to a few hours. The DuPont chemists who developed explosives and motion picture film found that by modifying the cellulose used in these products, they obtained a low viscosity lacquer resin that could be sprayed as a coating. Being a lacquer, this coating dried through solvent evaporation, rather than a chemical reaction. This new lacquer resin provided an excellent basis for a paint that had improved appearance, toughness, and durability compared to natural oil resin paints, and also could be pigmented easily with a wide variety of color pigments, besides just black! After a couple of years of testing, General Motors introduced *Duco* finishes on almost their entire automotive line in 1924.

But, as Paul Harvey used to say...here's the rest of the story. While several companies were working to develop a better automotive paint, DuPont sort of stumbled on theirs from a mistake. On a hot Friday afternoon in July of 1921, a DuPont worker left a drum of gun cotton (cotton fibers treated with a nitrate solution used in the manufacture of explosives) out on a loading dock during the very warm weekend. When workers brought in the drum on Monday morning and examined the contents, they found not gun cotton, but a clear, thick liquid. This liquid became the foundation of nitrocellulose lacquer, which soon became the leading automotive finish for nearly 50 years.

Alkyd enamel is another finish developed during this time period. It used a polyester resin modified by the addition of fatty acids and other components that made a more flexible coating. During the 1930s, both nitrocellulose lacquer and alkyd enamel became common. Many manufacturers also used combinations, offering black enameled fenders and a selection of colored lacquer bodies. Chrysler, Ford, Nash, Willy's and Studebaker used alkyd enamels, while General Motors and many smaller independent manufacturers used lacquer. These products and their derivations were used well into the '70s by both manufacturers and the refinishing industry.

Despite their popularity, these lacquer topcoats had one significant drawback – they had weak exterior durability. After about one to two years of exposure to sunlight and the elements, the coatings would begin to degrade, and frequent waxing was needed to restore the paint's shine. By the 1980s, the automotive manufacturers realized they needed better durability, as many consumers expected their cars to last at least five years, and they wanted their car to keep their new-car shine. At the same time, the Environmental Protection Agency was issuing new regulations for volatile organic compounds (VOCs) that limited the amount of solvent that an automotive facility could emit into the atmosphere.

Thus things changed again by the 1980s as European manufacturers demonstrated a superior product, the base coat/acrylic urethane clear coat system (Again, don't worry about the chemistry. It's just another way of making the binder.). Why? At the end of World War II, the Axis powers in Europe were prohibited from using certain chemicals and substances in their manufacturing processes because they also could be used to manufacture explosives. This put many of the components for nitrocellulose lacquer off limits. As a result, European manufacturers focused on enamels and developed a high-solids acrylic urethane paint system. This system was extremely durable in comparison to the air-dry systems of lacquer or synthetic enamel that were commonly used at the time. The other big advantage was that instead of the pigment being exposed to the elements, the pigment was sealed under a protective, transparent layer of clear coat. Paint manufacturers also began including UV inhibitors in the formula.

American manufacturers were slow to accept the European paint system, as the manufacturing and painting processes were more complex. Two problems associated with the new system were "dieback" and "mottling." Both were caused by successive layers of paint being applied before the previous layer had dried completely, thus trapping solvents in the lower layers of binder, and causing either loss of the glossy appearance, pigment particles clumping together, or both. With auto manufacturers sticking to their well-known paints, it fell on the auto repair industry to wring out the new system and eventually convince manufacturers of its superiority.

In the 1990s, another major development occurred in the formulation of automotive coatings: the use of water-based basecoats. The chemistries of these basecoats can vary between manufacturers, but the common factor is the use of water as one of the main volatile components. Typically, the motive to use waterborne technology is to obtain lower VOCs and reduce the environmental footprint of the painting process, but that is not the only benefit. It turns out that waterborne automotive basecoats often can provide improved appearance and metallic effects. *(Author's note – my 1987 Oldsmobile was a beautiful burgundy color, but the paint began to crackle into a spider web pattern after five or six years. In response to my complaint, Oldsmobile told me that water-based enamel does that, so get lost. Why was I not surprised?)*

I didn't get into it, but metallic paints, which became increasingly popular as designs evolved from straight lines to curves, were introduced in the 1930s. But there's more behind the story of metallic paints. Initially, these paints were made from fish scales. It would've taken about 40,000 herring to make one kilo of paint, but they'd give paints a mother of pearl look. For most folks, the expense of fish scale paint was out of the question. American paint companies substituted aluminum flakes, which were much cheaper than fish scales but this had their own set of issues. The metallic particles had to lay flat in the layer of paint to give the proper sparkle, and this introduced problems that the paint chemists had to overcome.

I also didn't get into advancements in primers, but have you noticed that cars don't seem to rust out like they used to? There also was a steady evolution in spray equipment, as manufacturers and repair shops needed spray guns that transferred most of the paint to the car instead of the air. And...there's also robotic spry equipment that coats consistent, even layers of paint without any runs, drips, or errors (and doesn't need coffee breaks, call in sick, pay raises, or go on strike).

The final major step change in automotive coatings technology occurred in the 2000s, and this advance focused on process efficiency. In the typical automotive assembly plant, the painting operation can take up to half the space of the entire facility, account for approximately 40% of the capital cost of an assembly plant, use 80% of the energy, and produce the vast majority of carbon dioxide and VOC emissions! OEM manufacturers have asked the paint industry to find ways to reduce this footprint and the cost of applying the coating. This required paint manufacturers to develop coatings that can be applied more efficiently, in fewer steps, and with less energy.

The paint and painting processes used today might seem obvious now, but look at it this way — without these kinds of innovations, cars couldn't be mass-produced and would've proven too expensive for most of us and thus, a whole lot more of us might still be riding bikes or walking. And it's certain that in the future, car paint technology will continue to evolve.

Sources:

1. https://www.paint.org/coatingstech-magazine/articles/brief-history-automotive-coatings-technology/
2. https://www.bodyshopbusiness.com/behind-the-paint-history/
3. http://islanddetailandcolor.com/a-brief-history-of-automotive-paint/
4. https://didyouknowcars.com/the-history-of-car-paint/
5. https://www.consumerreports.org/consumerist/a-brief-history-of-car-colors-and-why-are-we-so-boring-now/

MISCELANEOUS THINGS

The Cornell-Liberty Safety Car

You probably never heard of it, but this car influenced many safety features in our modern cars.

Back between the late 1940s and late 1960s, the Cornell Aeronautical Laboratory, part of Cornell University in New York, was *the* center for aerospace research by private industry. However, they also were involved in automotive and highway safety efforts, using aerospace technology to make driving safer and to save lives. Concerned with the rising accident and death rates on our highways, Cornell Lab partnered with the Liberty Mutual Insurance Company in the 1950s to develop and demonstrate many new safety features that were years ahead of their time. Cornell Lab was privatized in the late 60s, and still operates privately today, under the name of the Calspan Corporation, and you certainly know about Liberty, thanks to their advertising using a giant bird and this clueless guy with a toothpick in his mouth.

Over the first half of the 20th century, many automakers focused their efforts on making cars more reliable, more comfortable, and more powerful. Safety was a lesser concern. There were exceptions such as laminated windshield glass, which didn't break into sharp pieces, but back then safety didn't sell. For a while Ford tried hard to publicize the safety features in their cars, but it made little impact on sales. As Henry Ford II remarked: "We're selling safety, Chevrolet is selling cars."

The booming postwar economy put more Americans behind the wheel each year. Cars became faster and four-lane highways were being built to expedite people getting from here to there. It's little wonder that more drivers traveling at faster speeds led to a rise in accidents. By 1950, some 35,000 people were dying in auto accidents each year. Researchers at Cornell University and officials at Liberty Mutual insurance took notice. In 1951, the two institutions teamed up to research a simple question: What causes injuries in automobile accidents?

The Cornell-Liberty team made many discoveries by carefully studying accident reports and medical records from around the country. Car doors were a weak spot. Too often in an accident, a door was smashed open and one or more of the occupants was thrown from the vehicle. Furthermore, the team discovered that someone thrown from a car was more than twice as likely to receive serious injury, yet back seat passengers were three times safer than those in the front during a crash.

Researchers determined that the head was the most-frequently injured part of the body, and that one in ten victims received a facial disfigurement. Contrary to popular belief at the time, the steering wheel provided no extra protection to the driver. Indeed, the wheel was often a *cause* of injury, being pushed into a driver's chest during a crash. What's more, control knobs, window frames and decorative ornaments often maimed accident victims.

But the Cornell-Liberty team soon realized the auto industry and the public weren't being convinced by research findings published in reports and presented at automotive conferences. They needed to build a car to show how such features could be integrated into a stylish vehicle and how they would appear.

The Cornell-Liberty team put its findings into practice by building a concept car that reduced or eliminated many of these dangers. While many of these safety concepts seem obvious and are standard in today's cars, they were quite imaginative by mid-50s standards. Starting with a 1956 Ford Fairlane, the team incorporated more than 60 protective features. The safety car's accordion-style doors latched in three places, keeping them closed in a crash. Its energy-absorbing bumper wrapped completely around the vehicle, providing protection in low-speed accidents. Seat belts and bucket seats helped keep occupants in place in a crash, while head restraints reduced whiplash injuries. Dual roll bars integrated into the structure kept the passenger compartment from crushing in a roll over. Perhaps the most prominent difference was relocating the driver from the left side to the middle and replacing the steering wheel and column with a pair of control handles, making steering similar to pointing a sled. The dashboard, like other interior surfaces, was padded. Door handles, buttons, and knobs were recessed. Unnecessary trim and decorative pieces were eliminated.

Since avoiding an accident is better than surviving one, the Cornell-Liberty team took driver visibility and distraction into account. With the roll bar well-back of the normal roof support location, the wrap-around, distortion-free windshield gave the driver an unobstructed view, and was cleaned by five wiper blades. Also, the front passenger seats were set back a bit, so that passengers' heads wouldn't further obstruct the driver's vision. The oversized speedometer and gauges were placed directly below the driver's sightline making them quicker to view.

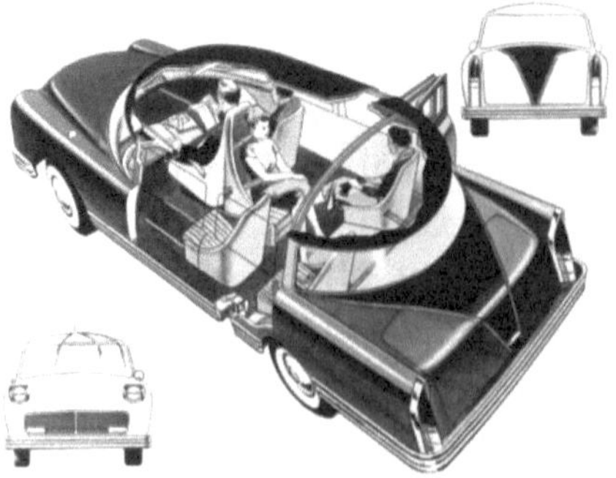

Neither Cornell nor Liberty Mutual had any plans to manufacture or sell cars. Instead, they hoped that their project would bring more attention to crash protection from the public and – more to the point – from automakers. A decade later, after some additional prodding from Ralph Nader and new government regulations, safety became an established priority. And while the car you drive today may not have five windshield wipers or handlebar steering, it certainly has a bit of the Cornell-Liberty Safety Car inside.

The car toured many cities and appeared at many auto shows, but I couldn't determine if it was street legal or if it actually ran. It eventually was donated to the Henry Ford Museum in Dearborn Michigan, where it can be seen today.

Thanks and apologies to the following for my blatant plagiarism of their articles:

Lou Knotts, Calspan Corporation CEO, from the book Calspan – Testing the Future

Matt Anderson, Curator of Transportation at the Henry Ford Museum.

David Burrell, editor of retroautos.com.au

Early Automotile Accidents

One of the unfortunate consequences of operating an automobile is the risk of an accident. It seems like there have been accidents ever since there have been cars. Even before then, there must have been horse-drawn carriage accidents. Indeed, researchers recently concluded that King Tutankhamun of Egypt died from a chariot running over him and breaking most of his ribs.

Cugnot's 1771 fardier à vapeur, as preserved at the Musée des Arts et Métiers, Paris.

One of the first self propelled vehicles was developed in France in by Nicolas-Joseph Cugnot (1725 – 1804). It was an awkward-looking, steam-powered vehicle. This claim is disputed by some sources, which suggest that Ferdinand Verbiest, a member of a Jesuit mission in China, may have built a steam-powered vehicle around 1672. But, it was too small to carry a driver or passengers. Cugnot was one of the first to employ successfully a device for converting the reciprocating motion of a steam piston into rotary motion by means of a ratchet arrangement. A small version of his three-wheeled *fardier à vapeur* ran in 1769. (A *fardier* was a massively built two-wheeled horse-drawn cart for transporting very heavy equipment such as cannon barrels).

The following year, a full-size version of the *fardier à vapeur* was built, specified to be able to carry 4 tons and cover 2 *lieues* (7.8 km or 4.8 miles) in one hour, a performance it never achieved in practice. The vehicle, which weighed about 2.5 tonnes empty, had two wheels at the rear and one in the front where the horses would normally have been. This front wheel supported the steam boiler and driving mechanism. The power unit was articulated to the "trailer" and steered from there by means of a double handle arrangement. One source states that it seated four passengers and moved at a speed of 2.25 miles per hour. The vehicle was reported to have been very unstable due to poor weight distribution - which would have been a serious disadvantage seeing that it was intended that the *fardier* should be able to traverse rough terrain and climb steep hills. Boiler performance was also particularly poor, even by the standards of the day, with the fire needing to be relit and steam raised again every quarter of an hour or so, considerably reducing overall speed.

The first "automobile accident"?

In 1771, Cugnot's's second steam-powered vehicle is said to have crashed into a wall during a test run, in what would have been the first automobile accident. However, it is disputed that this ever happened, since the earliest mention of this occurrence dates from 1801 and it was not featured in contemporary accounts. Cugnot's vehicle was acquired by the Conservatoire National des Arts et Métiers in Paris in 1800, where it can be seen today.

The world's roads apparently were safe from accidents involving powered vehicles for nearly another century. Then, in Ireland on August 31, 1869, Mary Ward was thrown from an experimental steam-powered car while riding as a passenger with cousins who'd built it. She fell under the wheels as it rounded a bend. This is believed to be the first recorded automobile death.

In 1891 John William Lambert of Ohio City, Ohio, was involved in the first automobile accident in American history. In the late nineteenth and early twentieth centuries, Ohio innovators in

Cleveland and elsewhere were at the forefront of this new form of transportation. Lambert was developing a vehicle in great secrecy. Operational tests were conducted at the farm implement showroom with the window blinds pulled shut. Outside road tests were conducted at night on little-used roads. That was a good thing since many times there were long intervals between being able to get the automobile started again after it stopped. It turned out that the villagers didn't even know what Lambert was working on. Lambert's vehicle—the first single-cylinder gasoline automobile, which was carrying Lambert and James Swoveland, hit a tree root, causing the car to careen out of control and smash into a hitching post. Injuries from this accident were minor. Lambert set a price of $550 in a sales brochure which was mailed during the first part of February of 1891. No sales contracts were signed for the Lambert, and Lambert soon realized that there was no sales potential for his automobile.

On August 17, 1896, Bridget Driscoll became the first pedestrian to die in a petrol-engined car accident in the United Kingdom. As she, along with her daughter and a friend, were crossing the grounds of the Crystal Palace in London, she was struck by an automobile belonging to the Anglo-French Motor Carriage Company that was being used to give demonstration rides. The accident happened just a few weeks after a new Act of Parliament had increased the speed limit for cars to 14 miles per hour, from 2 miles per hour in towns and 4 miles per hour in the countryside (the car's maximum speed was 8 miles per hour). Acording to witnesses of the time, Driscoll saw the vehicle "zigzagging" towards her and simply froze out of either fear or utter confusion, which resulted in her not getting out of the way. Following an inquest into the accident, the jury returned a verdict of "accidental death," and no prosecution was made. The coroner, Percy Morrison, said he hoped "such a thing would never happen again."

The first driver fatality happened in 1898, when Englishman Henry Lindfield and his son were driving from Brighton to London. Near the end of their trip, Lindfield lost control of the car while going down a hill. They crashed through a fence and Lindfield was thrown from the driver's seat before the car ran into a tree and caught his leg between them. His son was not hurt and ran for help. At the hospital, surgeons found the leg was crushed below the knee. Lindfield remained unconscious and died the following day.

Then, on September 13, 1899, Henry Hale Bliss became the first person killed by a motor vehicle accident in the United States. As he was disembarking from a streetcar at West 74th Street and Central Park West in New York City, an electric-powered taxicab struck him and crushed his head and chest. He died from his injuries the next morning. The driver of the taxicab, Arthur Smith, was arrested and charged with manslaughter but was acquitted on the grounds that it was unintentional. A plaque was dedicated at the site on September 13, 1999, to commemorate this event of a century earlier.

And, finally, what about the world's first crash involving two automobiles? According to popular legend, this occurred in Ohio in 1895, at which time there supposedly were only two automobiles in the entire state! This story never has been validated. Ohio didn't even require automobiles to be registered until 1905, so who knows how many there were a decade earlier. Additionally, the first sales of commercially available cars were made by the Duryeas Brothers in 1896. As such, it seems unlikely that one, let alone two, people owned motor vehicles in Ohio in 1895, unless they might have been experimental, home-made vehicles.

Author's note – Somewhere in editing I lost the list of sources for the above article. My apologies to those unknown sources.

An A Story

This isn't an A story as in MGA. It's a Ford Model A story. But most of us love reading about cars *and* airplanes, and this story includes both. So, here goes.

I first met my friend Terry Lutz about the mid 1970s. I was working at the flight simulation laboratory at Wright-Patterson Air Force Base in Dayton, putting together a simulation of the new F-16 fighter. Terry had just returned from his first flying assignment, flying F-4 Phantom IIs in Germany. We quickly became good friends. A few years later, Terry went off to attend the Air Force Test Pilot School at Edwards Air Force base, and eventually became an instructor there. By that time, I was managing my laboratory's flight test aircraft, so our paths crossed many times. Although Terry's work since then has taken him around the world, the internet has made it easy for us to remain friends (to tell you the truth, I think we've seen each other only once in the last twenty or so years). It was at Edwards where Terry bought his Model A.

By the mid 80s, Terry left the Air Force and went to work in Buffalo NY as a test pilot for the company that operated my aircraft, so we continued to see each other often. At his home for dinner one night, he showed me a load of parts he was cleaning, but they meant little to me, as it would still be many years until I would become a car guy.

By about 1990 or so, Terry moved again, this time going to Northwest Airlines as a DC-9 and A320 pilot. He then took an early retirement a few years later and moved to France to work for Airbus. How many people that you know spent the first day on a new job flying in the right seat of an A380? Is that cool or is that cool?

Terry's Model A before restoration.

Throughout his career, Terry continued to haul Model A pieces around the country and work on them a bit at a time. Here's the story as Terry tells it.

I purchasded my Model A from Hurley Hansen, who was one of the aircraft crew chiefs at the NASA Dryden Flight Research Facility. I met him when I was flying a test program on NASA's C-47 aircraft.

Here is what I started with in 1980. The original Sport Coupe top of the Model A had been removed and replaced with some sheet metal to protect against the sun and what little rain falls in the high desert. The rumble seat lid and panel below the rumble seat also had been removed so you could put a few bales of hay in the back, creating what you might call a "desert pick-up truck". When I got the car home, it became apparent that it had a 1930 body mounted on a 1928-29 frame, with 1928-29 fenders. I decided that the best choice was to keep the body, update the frame, replace the fenders, and restore it as a 1930 Sport Coupe. I worked on the A on and off, but mostly off, over the next three decades as time permitted.

What a difference 30+ years makes!!

By 1986, I had the car up through primer, after taking it all the way down to nothing on the frame, and building it back up to a car with doors and some old seats I bought at a swap meet. I overhauled the engine in my garage at Edwards, but never ran it. Almost everything I needed to finish the car, you know, the hard parts like lights and radiator shell and landau arms, were in boxes. In 1986, I moved to Buffalo to work at Calspan as an engineering test pilot (Steve Markman was our Air Force program manager, by the way), and no more work was done on the car. It moved from Buffalo back to Michigan when I left Calspan after a few years and it sat in a storage building near my house for the next 13 years while I was flying for Northwest Airlines. During that time, I built an RV-8 homebuilt, in the same building, after adding some heat! It remained there when I went to France to become a test pilot for Airbus. Parker, the guy that painted the airplane, runs a body shop, and is excellent at what he does. (*editor's note – somewhere in editing, I deleted Parker's first name...sorry about that.*)

I finally got serious about the A a few years later, after a hiatus of about 26 years. After living in France, I figured, what the heck, I'll talk to Parker about finishing the car. At that point it was still mainly paint and assembly work. So during the Fall of 2009, I was home and inventoried the boxes that contained all the parts. Parker trailored the car to his shop, and I took all the boxes. He worked on it part time, and we discussed this and that on email, and on short visits home. Each time I was home, we would discuss progress and what to do next. Parts needed to be ordered, such as tires and glass for the doors and rear window. At the end, he got it running and did some initial driving on the grass runway that he maintains. We had some starter problems and some water leaks, but so far no big problems. I really didn't get home that much, and let Parker work his magic. Eventually, he ran out of parts to install, and I drove it home!

It looks great, but still needs some work on the engine to get it running like I know it should.

A Few Anecdotes about Desert Car Lovers

Thanks again to my friend Terry Lutz for the following story. By the way, I have to thank Terry for suggesting the title of this book. It had nothing to do with cars, but Terry remembered that I always had Dachshunds, and that's how he remembered them.

In the late 1940s, the area around Edwards AFB, where I lived in the 1980s, was alive with old cars. I remember driving east of Lancaster on Avenue J and seeing a mile-long fence made with Model T frames stuck in the ground as fence posts.

I had a lot of help with my Model A from a guy named Don Ross. Don was a no-kidding rocket scientist, having worked for many years on Lehman Ridge overlooking Rogers Dry Lake, where rocket engines were test fired. Don restored Model A's in his garage in Lancaster, and did a great job. One of the many things he taught me was how to "derust" metal using a battery charger. I completely disassembled the body from my Model A and cleaned both the rust and the paint from all the sheet metal parts with this method.

At the end of each summer, Don would compete in a potato race at the Antelope County Fair. Five potatoes, painted red, where laid out in a line every 20 yards. The objective was for someone in the right seat to spear the potatoes by hanging out the right window, with a stick that had a nail in the end. The driver could only guess exactly where the potatoes were. I saw Don drive the course while some young guy speared all 5 potatoes in less than a minute.

Don was a college student during the Great Depression, and supported his family by hunting skunks and selling the pelts. He had enough extra to buy a Model T Ford to drive. That car had a special body, which made it an "Opera Coupe". Many years later, he found one of these rather rare cars, and restored it to like new condition. Sadly, Don donated it to a Museum in Missouri, which went bankrupt, and the car was sold at auction.

Another Antelope Valley old car restorer was John French. John had a Model T and each year he raced the Model T in a race called the Montana 500. This was a flat-out race against the clock, but there were some rules. The windshield was shut at the beginning of the race, secured by a tongue depressor. Eventually, the guys had their cars suped up enough that race officials made them put restrictor plates under the carburetors.

Terry Lutz with his RV-8 homebuilt.

John always had his Model T at the Antelope Valley Fair. He competed against other Model T's in a two-lap race. After the first lap, the car had to stop in front of the grandstand. The driver had to jump out, jack up the right rear wheel, remove it, run around the car with it, and then reinstall it before completing the second lap. He also raced against a horse each year, for one lap. The horse always won, to the great delight of the crowd.

Finally, I must mention Carl Bergman. Carl simply knew how things worked. He would go to a flea market and buy some obscure contraption, just to know how it worked. He had a card made up that said "I Like Old Cars!" If there was an obscure car out there, Carl had restored it, including a car once sold by Sears and Roebuck, featuring an umbrella for a top.

One year, Carl went to Nebraska on the rumor there was a Cretor's popcorn wagon mounted on a Model T frame that might be for sale. He found the popcorn wagon lying on its side in a dry wash. He made a deal with the owner, and gave him a check for $500 to hold it. He told the guy he would be back in 2 weeks with a trailer to pick it up. When Carl arrived with the trailer, the guy said

"Who are you?" When Carl reminded him about the deposit, the guy reached into his shirt pocket and said "Here's your damn check…". But Carl was the wiser one. He pulled out a wad of cash and started peeling off $50 dollar bills for the balance owed. Proof that money talks!

Carl restored that popcorn wagon to full functionality. Each year, he and his wife Alma would take it to the Antelope Valley Fair and hand out bags of popcorn to the kids. The last year I saw him, Carl had one of his gadgets going next to the popcorn wagon. He found a hand-cranked gadget that would make rope out of heavy string. With several large rolls of different colored string, he would make 6-foot lengths of rope for the kids, with the colors of their choice.

I could say a lot more about the old car people I met in the high desert. They were a wonderful bunch and they helped me with a little perspiration and a lot of inspiration. They were the kind of people you never forget. If they're no longer with us, I hope that others have come along to take their places.

My 1916 Wurlitzer Player Piano

This story doesn't really have anything to do with cars, but I thought you'd enjoy it anyway. I've often told people that I'd always wanted an MG TD. Not that I'm especially a car guy, but I just always wanted a TD. Well, there's something else I always wanted: a player piano – not that I love pianos or even know how to play one – I just always wanted one.

Player pianos were made by the tens of thousands in the first three decades of the 20th century. If you wanted live music in your house and your wife or daughter didn't know how to play a piano, a player piano was the solution. With the passage of time and improvements in sound quality, people eventually turned to radio and phonographs. Players aren't exactly rare today, but many of the ones that have survived in homes don't work. They're complex to maintain, and expensive to restore. Paper rolls are readily available at antique markets and especially on e-bay, and one company still makes new ones, including many current titles.

At an MG Car Club meeting a few years ago I was chatting with one of our members. He was in the final stages of finishing his basement and had one corner to go. But he had a problem. There was this player piano sitting there. He'd bought it about ten years earlier intending to restore it. He'd started, having disassembled it and refinished the case. For some unknown reason his interest turned (or returned) to restoring cars and he never got around to rebuilding the various mechanism's that make the piano work. "I'd give that thing away if I could ever find someone dumb enough to take it in this condition." Enough said about how it quickly changed owners. He even delivered it to my house.

The piano sat in my garage for a few weeks while I scratched my head trying to figure out how to get it into my basement. I finally bit the bullet and hired a piano mover. Five guys who looked like former NFL players came out, lifted it on to a special skid, and in fifteen minutes had it in my basement and were gone. That was the first $250 expense.

My friend seriously intended to restore the piano; many of the smaller parts were in bags or boxes and labeled. There even were bags and bags of replacement parts. He assured me all the original parts were there. The down side was that he hadn't taken any pictures as he disassembled the piano, since he'd intended to put it all back together soon. So, here I had this big jigsaw puzzle, with no idea of what went where, what worked, or even what the various assemblies were supposed to look like or function if they worked.

I found a local piano repairman who worked on players. He told me that while Wurlitzer was best known for its big theater organs, their pianos were more middle-of-the-road, sort of the Chevrolet of pianos. He looked up the serial number in a reference book and determined that it was built in 1916. Written in pencil on the cast iron harp was the name H. Shaffer and the date 11/9/27. He told me that it was common practice for repairman and tuners to do that when they worked on a piano.

The piano sat in my basement for another two years until it finally surfaced at the top of my project list. A player essentially is a standard piano with the player action (the mechanism that reads the holes on a paper roll and actuates a small bellow to lift each hammer on the piano action), and a foot-operated pump to produce the vacuum that drives the player action. The player action appeared to be the most complex assembly, and in the worst shape. It had eighty-eight bellows, one for each key, and the fabric was falling apart on about half of them. I decided to tackle this part first.

I looked on the internet for a source of information and materials and stumbled upon a place in Cincinnati called the Player Piano Shop. I gave them a call, then put the player action in the car and drove it to Cincinnati for a consultation. The owner turned out to be a guy named Steve who loved player pianos as much as most of us love our MGs. He's been in the business for about forty years, learned it from the previous owner of the shop, and to make it short, knows every detail of every brand of player piano ever made (there were dozens and dozens of them). Walking into his shop is like walking into an old-time store: sagging wood floors, peeling paint and wallpaper, piano parts stacked everywhere, drawer after drawer of small parts (both salvaged originals and modern replacement items), and not a computer monitor in sight. The most modern item there was a Princess phone, but no answering machine attached to it.

Steve looked over the player action and said that the fabric on all eighty-eight bellows needed replacement. He removed one, showed me how to do it, and then gave me enough rubberized fabric to do the remaining eighty-seven bellows. I spent two hours there and Steve just couldn't stop talking about player pianos and how happy he was to have met someone else interested in rebuilding one. We quickly became friends. It took me about three months to remove, recover, and reassemble the eighty-eight bellows and remove and overhaul the valve that went with each one. I then spent another few weeks recovering the bellows on the tracking mechanism (which keeps the paper roll centered) and the vacuum motor that drives the paper roll.

The next step was the piano action, which consists of (you guessed it) eighty-eight hammers and the articulating mechanism for each that causes the hammers to strike and instantly pull away from the string. As an experienced woodworker, I can attest that these pieces were the most beautifully made wood pieces I'd ever seen. Those piano makers from long ago knew how to select just the right type and piece of wood with the right grain to give strength to last for centuries (I found only two pieces with slight cracks, which I repaired. Btw – these are made of plastic on most new pianos today). I removed each mechanism, cleaned it, and sanded smooth the felt tip on each hammer. After reassembly, I placed the piano action into the piano. I then took the keys, cleaned them, and installed them after having Steve replace the ivory key tops on about a dozen of them (btw – it's illegal to buy or sell new ivory, so you have to salvage the tops from old pianos).

Next, I installed the player action and moved on to the pump, which consists of two large bellows with leather flapper valves and two spring-loaded bellows that serve as reservoirs and to smoothen out pulsation from the pump. Again, Steve gave me the correct bellow cloth and leather. These bellows were big, but the process for covering them was the same.

With the pump finished, I installed it, hooked up the new rubber tubing, then sorted out the pieces for the pedals and determined where they went. Steve came up from Cincinnati and did some adjustments to make sure everything was tweaked just right. As soon as this was done, I inserted a roll (I'd been buying them on eBay for about a year, and had a nice collection of mostly 1920s, 30s, and 40s music). It sounded awful, but definitely was recognizable as Chattanooga Choo Choo (what I should have expected for a piano that probably hadn't been tuned for at least two decades and had been moved who knows how many times?).

I worked for another month or so, sorting out the final small pieces and the remaining pieces of the case, then had the piano tuned. What can I say? Pure pleasure, just as much fun a driving my MG TD (but no concern for breaking down on a lonely country road).

Note – Throughout this effort, I tried to make as correct a restoration as possible. All screws that had to be replaced were replaced with ones with slotted heads, which are getting hard to find, and in general don't seem to have quite the same shape on the head nor the same plating. While the original used only hot melt hide glue, I gave in to modern convenience and used liquid hide glue, the modern alternative. In gluing the bellows, I used Titebond III, per recommendation by Steve, as it is easier to use, less messy, and is resistant to the long-term effects of humidity. For the little bit of surface finishing that I did, I lacquered with Deft, the modern version of cellulose nitrate, which is what was used back then (Use only in a well-ventilated work area, as the fumes will knock you out – don't ask how I know that.).

What the Heck is a Mazda

I'm not talking about a car

I trust you enjoyed reading about my 1916 Wurlitzer player piano, even though it wasn't about a car. Now, I'll talk about a Mazda, but it's not a car, and my search to find out what it is, and why it's in a 1952 song.

One of the paper piano rolls I bought on eBay is the Glow Worm song. It's always been one of my favorites; the tempo, words, and rhyme seem to flow effortlessly, despite the lyrics' continuing references to light and electricity not making a whole lot of sense. The song was written by lyricist and singer Johnny Mercer (1909-1976). While Johnny and others recorded it, the best known recording was by the Mills Brothers. However, Mercer actually adapted the melody from a German song called "Das Glühwürmchen", a song from Paul Lincke's 1902 operetta *Lysistrata.* The original lyrics talked about worms that light a path in the forest for young lovers to follow. Mercer kept only the melody from the chorus, and made it about fireflies rather than worms. But enough about the background of the song.

In the last verse are two lines that state:

> "You have a cute vest-pocket Mazda
> That you can make both slow and "fazda."

I'd heard the song many times over the years but payed little attention to the lyrics. But, singing along with the words printed on the paper piano roll, the word "Mazda" caught my attention. What were they talking about? Even if Mazdas were around in 1952, they weren't being sold in the United States so who ever would have heard the name? I looked up the word in the dictionary and found it had something in the Buddhist religion to do with enlightenment. Made some sense, but what is a "vest-pocket Mazda," and why was it capitalized?

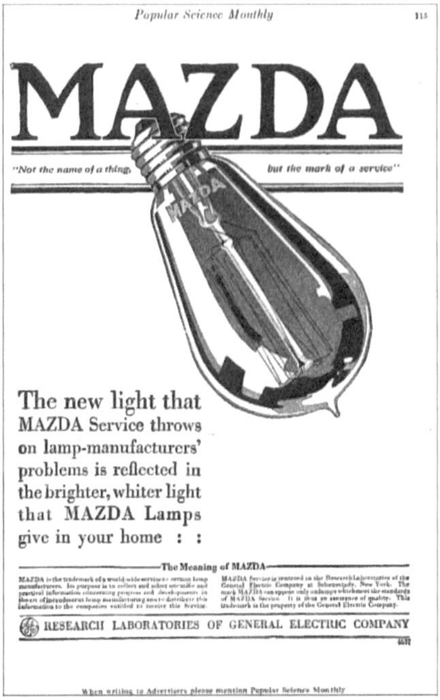

I then turned to the all-knowing source of information, Google, and typed in "vest pocket Mazda." The mystery slowly began to unravel.

First, I found out it was a light bulb. Mazda was a trademarked name registered by General Electric in 1909 for incandescent light bulbs. The name was created for bulbs made with a tungsten filament. Up until this point, light bulbs used carbon filaments. They were inconsistent in brightness and burned out quickly. Tungsten filaments gave consistent brightness and long life. GE licensed the Mazda name, socket sizes, and tungsten filament technology to other manufacturers to establish a standard for lighting. Bulbs were soon sold by many manufacturers with the Mazda name licensed from GE, including British Thomson-Houston in the United Kingdom, Toshiba in Japan, and even to GE's chief competitor, Westinghouse. Tungsten bulbs of the Mazda type initially were more costly than carbon filament bulbs, but were far superior.

GE's patents on the tungsten filament lamp expired in the late 1930s and they stopped using the name by 1945. They also stopped licensing the trademark to other manufacturers, although it continued to renew the trademark registration up until 1990. BTW -

ask anyone about ninety or so if they ever heard of a light bulb referred to as a Mazda lamp. If they remember, they'll probably say something like "I haven't heard of them called that since I was a kid."

So, I've firmly established what a Mazda is, but what is a vest-pocket Mazda and why would it go slow or fast instead of dim or bright? In my Google research, I stumbled upon a website called oldchristmastreelights.com. An e-mail to them got an explanation of the vest-pocket part:

> "80 years ago, when somebody made something about the size of a pocket watch or cigarette lighter, they said it was "vest-pocket" sized. Eveready, who made Mazda lamps, once advertised a "vest pocket flashlight."

Next, I stumbled on the Johnny Mercer Music Program, located at Braille Institute, and founded in the 1960s by Johnny and his wife Ginger, to help blind and visually impaired children and adults learn to sing and perform. They couldn't explain the lyrics, but referred me to Kevin Fleming, the Popular Music and Culture Archivist at the Georgia State University Library. Perhaps they would have some of Mercer's notes about the song. I contacted Kevin and he explained most of the remaining mystery:

> "You are correct, the vest pocket Mazda refers to the GE brand lightbulbs. The draft lyrics for the song has "fasta" in the next line, but the published sheet music has it as "fazda." The draft lyrics do not indicate what this means. However, the word "fazda" in the published sheet music has quotes around it, which I believe is an indication

to the performer that this word is deliberately made-up/misspelled."

"So, I would agree that "slow or fazda" is an alternative for dim or bright, and was used because of the rhyming scheme. Additionally, it fits with the car references in the same verse (e.g. "turn the key" and "taillight")."

Car reference? Suddenly a light bulb glowed (a Mazda bulb, of course). I'd totally overlooked the preceding two lines:

"Glow little glow-worm, turn the key on
You are equipped with taillight neon."

Johnny Mercer must have been a car enthusiast!! He made a reference to cars and was just continuing the comparison! The key ignition switch on cars was introduced in the late 40s, and neon lights were invented in the early 1900s, so the references to them are not inappropriate.

So, at this point, I'm satisfied with the meaning of the lyrics, although they're a bit nonsensical. But, two questions still remain.

1. Was Johnny Mercer a gearhead?
2. What symbol did cartoonists use to show a brilliant idea before light bulbs were invented?

I'll worry about these some other time.

If, by this point, I haven't gotten you totally turned off to the topic, you can go to Youtube and find recordings of the Mills Brothers and Johnny Mercer singing the Glow Worm song, and even a recording of Das Glühwürmchen.

My 1963 REO Snow Blower

I know!! It's not a car, either, but it has a gasoline engine and tires, and has its origins with Ransom Eli Olds, founder of Oldsmobile, so I'll share the story with you. You may recall the article about horsepower, and my being curious why my new Troy Bilt snow blower only listed the engine displacement and not the horsepower. Here's the story about the old REO snow blower that I replaced after nearly fifty years of service.

Way back in high school, besides having a paper route to earn money, I also mowed lawns and shoveled snow. Since we didn't have video games, computers, smart phones, and only three channels of television, I had plenty of free time and these money-making activities kept me busy. Snow blowers were a relatively new invention in the early 60s, and I kept bugging my father to buy one. My argument was that I'd be happy to plow the drive instead of being forced to shovel by hand, and I even could earn extra money clearing neighbors' drives. By about 1963, my dad finally relented and agreed to buy one, but I had to pay him back half of the cost, about $65, from the money I'd earn with it.

Our new REO snow blower was a bit of a beast, and lacked anything close to modern safety features. In fact, the gear selector was spring loaded in the forward position, so heaven help you if you fell and let go of the machine. Even while in neutral, a slight bump to the selector lever could pop it into gear and the machine would lurch forward. I suppose this was typical of the machine design philosophy of those days…the operator was supposed to pay attention to what he/she was doing and be careful. What an antiquated thought. The only thing close to safety devices were the sheer pins that fixed the numerous gears to their shafts. In the event the snow blower hit a rock or the corner of a house, the pin would snap, preventing further damage to the mechanism. Note that this was to protect the machine, not the operator or the house. More on this later.

I don't recall if we got it in late 62 or early 63, but with the heavy snowfall we used to get in Cleveland, I was plowing a few driveways a week at $3 per drive, a lot of money for a kid back then. It was a thrill eating through six or eight inches of snow and seeing the plume shoot twenty feet off to the side. Interestingly, the deeper the snow, the farther away and higher into the air it'd shoot. Even if the snow plow cleared our street and threw a mix of compacted snow/ice onto the drive before I was out there, the REO would eat right through it.

After a month or so, we had a major blizzard and schools were closed, a fairly rare event back then. I plowed my drive, and then got calls from a few neighbors. While I was out working, a friend wondered by. I offered him a job; go house-to-house and find me more customers. I gave him fifty cents for each one he found. Soon he came back with a list addresses. I was out all afternoon and into the early evening.

We always ate dinner together as a family. When I didn't come home, I don't know if my parents were more worried about me or more angry that I was absent. I finally came in as my parents were washing the dishes (by hand…no dishwashers in those days) and they started scolding me for missing dinner. As they yelled, I put my hand in my pocket and started pulling out crumpled $1s and $5s and making a stack on the kitchen table…sixty five of them to be exact, and said that this should pay off my half. I still remember the look on my Dad's face, one of the few times he was speechless! Dad lived for another fifty four years, and I don't think I ever saw that look on his face again.

My son clearing our driveway, 1992.

After I left home, my Dad continued using the REO until sometime in the late 70s, when he bought one of those little light-weight ones with an electric start because the REO was getting too hard for him to use. I brought it here to Dayton and continued to use it for years. It ran all right, but seemed to be loosing power over the years and was getting harder to start. I also was lucky to get a half-hour of running time before a sheer pin would break. This required about a half-hour to repair since I first had to remove the sheet-metal body. As mentioned earlier, the gears were connected to the shafts by shear pins. Disassembly revealed that the holes were getting warn so that there was movement between the gears and shafts, stretching the pins until they broke. The pins were supposed to snap to prevent damage, but they weren't supposed to break under normal use…almost every time I used the machine. I eventually replaced them with bolts, and hoped that I never hit the corner of the house or garage, as I'd probably do significant damage. Since REO had gone out of business years earlier, as had the company that made Clinton engines, parts weren't available and having them custom made was out of the question. While replacing the pins with bolts helped, still it seemed like I was spending more time tearing the machine apart than actually plowing snow. It also took forever to start, even after shooting starting fluid into the carburetor and using an electric space heater to warm the engine. By about 2012 or so, the rewind spring for the pull start cord broke, and I had it. I bought the Troy Bilt snow blower.

After all those years of service, I hated the thought of putting the REO out in the trash, and pushed it into the corner of the garage. Maybe I could find someone who'd want it. I thought that perhaps there was a museum of yard equipment somewhere. I Googled the topic every so often, and after a few years, I stumbled on someone in the Findley area, north of Dayton, who collected REO lawn mowers. I contacted him and he never knew that REO made a snow blower. He'd be more than happy to take it. A few days later he drove down to pick it up. I guess I shouldn't have been surprised that just as we love our MGs and take them to shows, there are yard equipment shows and this gentleman specialized in REO equipment. He showed me a picture album of all the mowers he's restored and pictures of him at shows with some of his prize entries. He was especially excited when I gave him the original owner's manual, one 11x17 inch sheet folded in quarters, with none of the typical safety warnings or legal disclaimers, and all in English. It even had the original sales receipt stapled to it.

As he was getting ready to leave, I realized the man's wife had been sitting quietly in the truck the whole time, so I went over to say hello. She smiled, shook her head, and said that everyone has to have their toys. He drove off happier than some of our MG Club members with a new, rusted junkyard MG to restore. I contacted the new owner a few years later and he still had the REO, but hadn't worked on it yet. Hopefully, some day it'll impress his fellow yard equipment enthusiasts, and maybe it will even take an award at a yard equipment show.

Who Invented the Phillips Screw?

While most screw heads are called something related to their shape, i.e., slotted, star, hex, I always wondered why the crosshead pattern was called a Phillips screw. Once again, the source of all information, the internet and Google, came to the rescue. Believe it or not, the Phillips screw was not invented by someone named Phillips, but a man named Henry Phillips had a lot to do with making it a success, so here's the rest of the story.

Metal screws were first used during the middle ages, and I'd imagine that each was made my hand and therefore quite expensive. The slot was the standard head design, as it was easy to make, as well as the screw driver with which to drive the screw. With industrialization in the late 1700s, methods of mass-production were developed. In 1770, English instrument maker, Jesse Ramsden (1735-1800) invented the first satisfactory screw-cutting lathe, and Englishmen Henry Maudslay (1771-1831) invented a large screw-cutting lathe that made it possible to mass-produce accurately sized screws. Then, in the early twentieth century, especially in the auto industry, mass production and the production line led to the development of power drivers for inserting screws quickly. However, anyone who ever used a power driver to drive a slotted screw quickly realized the difficulty to keep the driver centered and not damage the surrounding surface.

In the 1930s, along came an inventor named John P. Thompsen of Portland, Oregon, who developed a cruciform-shaped head and the matching screw driver. He must not have been a very good salesman, as he couldn't interest the expanding automotive industry in his invention (apparently not uncommon, as there were earlier cruciform head designs that never caught on, either). It is not known how Thompsen knew Henry F. Phillips, who was the managing director of the Oregon Copper Company, a mining outfit in eastern Oregon, but when the patents were granted in 1933, the rights were assigned directly to Phillips, even though Thompson was credited with the invention.

Phillips then formed the Phillips Screw Company, improved the design, and promoted the adoption of his product. Interestingly, the Phillips Screw Company in Portland never manufactured screws; its aim was to license the design to manufacturers and collect the royalties. He soon persuaded the American Screw Company to manufacture the screw, and in the next four years the Phillips Screw Company obtained six additional patents modifying the design. By 1936 the screw was available to consumers, and the first industrial customer was General Motors, which used Phillips screws to build Cadillac automobiles in 1937. Soon after, it was adopted by the railroad and aviation industries.

Henry Philips died in 1950 at the age of sixty-eight.

Microfiber Towels

Dry off your car the right way.

After washing your car, you need to use the right towel to wipe away excess water and soap. Even if you use the finest wash products, using a bathroom towel to dry your vehicle could risk scratching the vehicle's paint. Prevent this by using a microfiber towel.

What is a microfiber towel? It's a piece of cloth made of polyamide, polyester, or a combination of these fabrics. Each fiber used in the towel is smaller than a single silk strand and about 1/100th the diameter of a human hair. This gives a microfiber towel its extreme durability, preventing it from shrinking, wrinkling, or stretching after repeated use. The polyester used in a microfiber towel also allows it to absorb lots of water yet dry quickly. These properties also help prevent the formation of mildew, as the towel never remains wet for long.

But their real advantage is preventing abrasion to your car's painted surfaces. As their name suggests, their "micro" fibers are so tiny that they can grip and lift dirt into the towel away from the surface of the vehicle. Regular towels made of materials like cotton typically will just spread dirt around when they are wiped across the surface of a car. In addition, when a microfiber cloth's fibers rub together, they create a static charge. The static charge improves the cloth's ability to clean even more, because the charge attracts tiny dirt particles. The more polyamide in the blend, the softer the towel will be making it more appropriate for sensitive surfaces like your car's paint. Not only are the towels not abrasive themselves, they also lift dirt away from the surface. This reduces the chance of debris scratching the surface as the towel is wiped across the vehicle.

There are several features to consider in a microfiber towel, including:

Pile height: Microfiber towels can come in short, medium, and long pile heights. The shorter the pile, the more all-purpose the towel. You'll notice that short-pile towels are more rigid, while long-pile towels are soft and plush. Generally, you want to use a short pile on non-sensitive surfaces like dashboards or plastic trims, a medium pile for applying or removing car wax, and a long pile for buffing and drying your vehicle.

Material ratio: As previously mentioned, microfiber towels are usually a combination of polyester and polyamide. The best microfiber towels are 80% polyester and 20% polyamide. This combination makes the towel super soft while still absorbent. Towels that have a higher amount of polyamide may be more absorbent but less soft. But, a super high polyester towel may produce more lint.

Texture: Some microfiber towels are waffle weave, while others are suede smooth. Waffle weave towels are more absorbent and better to use when drying, while suede smooth ones are better for polishing.

Before you buy, consider your planned usage for a microfiber towel. For example, if your main goal is to polish and buff your car, you might want a suede smooth towel with a long pile. However, if you're interested in detailing or drying off your plastic trim, try a short-pile towel with a waffle weave texture. You may want to keep different types of towels on hand.

Finally, think about buying your microfiber towels in bulk. This way, you'll always have a clean one ready to go—and it's usually cheaper.

Sources: Truecar.com and Automotivetrainingcentre.com

What Is Human Factors and What Does It Have to Do With Cars?

Human factors (HF) is the study of how people use and interact with technology. It is the relationship of human abilities and limitations with any mechanical object or system, and how to design the system so that the operator can use it more effectively. Human Factors isn't an exact science, since it involves people, whose actions vary and are hard to predict precisely. However, there are general patterns that people tend to follow with some consistency. As a result, engineers and designers consider HF (or should, anyway) to determine how well a user can use a product and to modify the product as needed before it goes on the market. For example, if a machine requires super-human strength to operate (like my snow thrower), then the average person will find the machine unusable. Users might even think themselves failures, since they can't get the machine to work the way it should, or at best it is very difficult to use (again, like my snow thrower), when in reality, the problem is with the machine.

I'm not an expert in the subject, but have a long history with it. I worked with many human factors experts when I managed flight research programs for the Air Force Research Laboratory. I even took a human factors engineering course in college, taught by one of the best-known researchers in the field. It was one of the most practical courses I ever took! Seldom does a day go by that I don't use, or at least think about, one of the things I learned in that class or from other experts (example – why does your wrist sometimes get sore when carrying one of those little baby carriers, but not when carrying a heavy suitcase?).

One of the objectives of using human factors is to put all controls and instruments in logical arrangements that are easy to remember and use. Another is to make a machine comfortable to use. I'll give you a few examples. First is the heater and a/c controls on my 1997 Olds Silhouette mini-van. There were three dials on the dash, one each for fan speed, temperature, and heat & a/c modes. They were labeled clearly; I could see them with my peripheral vision, and I could grasp each one easily and operate each by feel while keeping my eyes on the road. Due to the shape of the knob I could tell just by feel where it was set. This is called "tactile feedback." Have you noticed that most controls on newer cars are buried in a maze of push buttons with the settings displayed on a video screen? So, you have to look down to put your finger on the proper button, then look at the video screen to see how you've adjusted the setting. Think about it...with no tactile feedback to help, you had to take your eyes off the road while you were doing this! And...can you push the buttons while wearing gloves? I'd take the three dials from my Silhouette any day over the clean look of all the push buttons.

My other example is the sand blaster I used when I volunteered at the restoration shop at the Air Force Museum. We had a vintage machine, probably made during the 40s or 50s. It was a big steel cube about four feet on each edge, mounted on a truss stand, with a window to see in and openings into which the user inserted his hands into heavy rubber gloves. It worked great, blasting rust and old paint off of metal parts. However, on larger parts, when I'd be standing at the machine for more than fifteen minutes or so, I'd step back and my shoulders and back would hurt. I assumed that that was the way it was, and the user just had to put up with it. Then, we got a second machine, a new one. The first time I used it I felt none of the discomfort of the older one. As a naturally-curious engineer, I stood back and compared the two. On the old machine, the side where the operator stood was vertical and the operator had to stand straight as a board. On the new machine, the operator's position was slanted inward just a bit, starting at the operator's waist, so that the operator was leaning slightly forward. With that barely-noticeable change, I could blast away for an hour without getting sore! The operator no longer had to put up with a poor design, as the newer design took into account the structure and limits of the human operator, thus making the operator a more effective user of the machine.

(BTW—the issue with the baby carrier has to do with how you position your wrist. With a suitcase, your fingers face your body, which is its natural position. With the baby carrier, your wrist is turned ninety degrees, so your fingers are facing either forward of aft, contrary to how your arm/wrist joint is made, causing stress, fatigue, and even pain.)

Another aspect of HF is called "population stereotype." Basically, these are things that are done a certain way, but for no special reason other than that's the way they've always been done. Did you ever think about the colors on the traffic signal? Is there something intrinsic in the use of green, yellow, and red meaning what they mean? Not really. What about the hot water handles always being on the left and cold water on the right? And what about light switches...why is 'up' on and 'down' off? And clockwise to tighten a bolt or drive a screw? And, something I'd bet money you never noticed...look on your push-button phone and note that the numbers 1, 2 and 3 are on the top row and rows of subsequent numbers are below. Now, look at your calculator. The numbers 1, 2, and 3 are on the bottom and the rows of subsequent numbers are above.

You went and looked, didn't you, and I'll bet you never noticed it before!!

In the aviation business, not only do we expect more and more from the pilot, but for him or her to be able to do things quicker and with fewer chances of making errors. If you look at the throttle and control stick or wheel on a modern military or commercial aircraft, you'll notice buttons and dials all over them. This concept was named HOTAS, which stands for Hands on Stick and Throttle. This placement of controls allowed the pilot to control all of the aircraft's critical systems without having to reach for anything. Of course, all buttons had to be carefully placed to be easy to use while maintaining a comfortable grip. Then, they developed the HUD, or Head Up Display, so that the pilot could see all essential information for the task at hand through the front windscreen, without having to look down at the instrument panel. The latest evolution of this is the helmet-mounted display, which projects critical information onto the pilot's visor, so that the pilot still can see essential information when looking off to the side.

Sometime back in the mid 70s we were doing a lot of research into new cockpit designs to make the pilots' job easier (or, to allow them to do more work without being overloaded). One day we hosted a tour for a group of automotive engineers who wanted to learn how we were using human factors to design better aircraft cockpits. Naturally, they were very impressed, but one of them pointed out the problem with implementing similar concepts in cars. In aviation, we can sit pilots in a classroom to teach them how to use the systems, and then put them in a simulator to gain experience. Finally, they can fly with an instructor in the cockpit to make sure they know how to use everything. They explained that this won't work for cars, as most car buyers would never sit through a class or even watch a video, and will just hop into their new car and drive off without even looking at the owner's manual. Therefore, anything in a new car has to be 100% intuitive with the driver receiving no instruction. The industry feared lawsuits by an injured driver who didn't use some new feature properly and caused a crash.

Fortunately, automotive engineers have figured out how to incorporate many safety features that basically are passive from the driver's point of view. Starting thirty or so years ago, some controls, such as those for cruise control, windshield wipers, lighting, and audio system started moving from the dash board to the steering wheel and/or turn signal lever, helping the driver to keep both hands on the wheel and their eyes on the road. More recently, self-driving car research has given us spin-offs such as blind spot and lane drift warning, adaptive speed control, and auto breaking. These all were inspired by decades-old cockpit research, and require no special training for the driver. . So, they *did* learn something from that tour!

So, what was the event that prompted me to write all this? My car was in the shop a while ago and I rented a car for a few days. Being frugal, I requested a small car, but they were out and gave me a Cadillac XTS. Don't get me wrong…it was a very comfortable car, but the dash was all push buttons and a touch-sensitive screen, and that's not the half of it. What looked like buttons weren't buttons at all; I couldn't get any of them to work as expected until I finally realized they were touch sensitive – don't push, just tap each one lightly to activate. To make matters worse, all were labeled with pictures: a few were obvious, but most just looked like an arrangement of randomly-placed lines, circles, and arcs. After a half hour or so of driving, I finally figured out how to set the radio and temperature. Fortunately I never had to use the windshield wipers. I never could figure out the cruise control, even though there was a button whose hieroglyph looked like something that represented speed. I can't imagine a Cadillac not having cruise control, so I know it had to be hidden there, somewhere. I'd have looked in the owner's manual, but it wasn't in the car.

And that snow thrower that requires super-human skills? I could have bought the next bigger model that had ratcheting wheels to make turning much easier. But it was about $300 more and two inches wider, so wouldn't fit between my cars in the garage. So, I just put up with it. I utter a few inappropriate words when I use it, but first make sure no children are within hearing distance.

And don't get me started about my gripe with baby strollers. (Hint – the front wheels caster while the rear ones point straight ahead, and the handle usually is very close above the rear wheels. It gives terrible leverage for steering, and I usually get a sore back after pushing one for even a short while. Why don't they put the castering wheels on the back?)

Where Have All the Children Gone?

I grew up in a post WWII plat where a new house was started every week. We kids would walk or ride our bikes to each new construction site, then sit and watch for hours. When the workmen went home at night, we'd come back. Although we had a public park and schoolyard nearby, these construction sites were our real playgrounds, and the earth-moving machines were our jungle gyms. I never wanted to grow up to become an engineer; I wanted to drive a bulldozer.

Jumping ahead a few years (O.K., a whole bunch of years), we moved into our new house in 2008 at the height of the housing bust. Soon after, the builder backed out on the development, and as a result, the street planned to be behind us didn't get built for several more years. During this time, we enjoyed the large, open field behind us and the clear view of the forest beyond that. That changed after a few years. The housing market improved and a new developer came in to finish building the plat. There was much construction activity just behind my fence.

Now, as an adult, I had a ringside seat watching the earth movers and bulldozers relocating tons and tons of dirt to grade the field for streets, and then watching the power shovels dig basements. I'd sit on my deck with a cup of hot tea in the morning or a cold drink in the afternoon and watch those giant machines hour after hour. Sometimes the operators would wave when they saw me watching, and I'd wave back. To a non-sports fan, this was like sitting on the fifty-yard line at Super Bowl. Unfortunately, I also watched many of the trees being cut down with factory-like efficiency.

In the evening I'd go and walk around those steel monsters, still amazed at their mass and solidness and raw power. Then, one day, I noticed something; I was the only one watching. Where were the kids? Although we're the senior residents in our neighborhood, most of our neighbors are younger and have kids. We see them outside playing on occasion when it isn't too hot, but there never was a kid outside watching the heavy equipment, never.

I guess they're indoors in their air-conditioned houses, watching one of their 100+ channels of cable, or watching Netflix, or playing video games. Maybe they're at one of those organized activities to which mom's haul their kid. Heaven forbid if they go outside on their own, without sunscreen, or get dirty, or play by the nearby creek or too far from the grassy park. What if their cell phone signal is blocked and they can't communicate with home? What if they're with the "bad" kid in the neighborhood? Will they get kidnapped if they're out of sight of their house?

Don't kids learn certain social and leadership skills by being allowed to play on their own, to explore to their own level of curiosity, to investigate new situations and decide for themselves what is safe and what is not? Today we call this a benefit/risk analysis. I don't think they can learn this process with their face buried in a video screen, where the only consequence is loosing a few points, or when a parent makes every decision for them. Somehow, we knew to keep a safe distance from those bulldozers, and sometimes we stumbled and slid down the mountain of dirt we'd just climbed. We also knew not to get too close to the hole that had just been excavated in the ground.

Is being out of touch with home a big signal flag to kidnappers? We knew to be home in time for dinner, then to be home before bedtime. I never had a watch until I was eleven or twelve, and don't recall ever being home late. Our Mom's knew we'd be home on time, and didn't worry about us, or who we were with. Do kids today even know what daydreaming is? That's what we'd do when we had nothing else to do and we'd explore an imaginary galaxy in our mind, or stand on the pitcher's mound at the World Series, or be a princess walking into the ball with her prince. You can't do that with your face buried in a video screen.

Did playing in the dirt somehow even keep us healthy? I've read articles about how kids seem to be much sicker now than they were a generation or two ago. Are we getting weaker, or when we were kids did we just keep on playing even when we felt sick out of fear that Mom would drag us to the doctor? Many animals, when they're injured, will roll in the dirt. Is there some antibiotic in the dirt that animals know about and we don't? Do kids today spend too much time indoors, where they're exposed continuously to all sorts of chemicals given off by carpeting, flooring, furniture, etc, that are trapped in a house in which the doors and windows always are closed?

If you think about it, you do a benefit/risk analysis every time you drive your MG or other classic car. You ask yourself "Is the pleasure I get worth the risk I'm taking?" Our classic cars feature few of the safety features of modern ones: no air bags, anti-lock brakes, etc. Some of them don't even have seat belts! Maybe you drive more cautiously and keep a better distance from other cars. Maybe you learned the process by sliding from the top of that dirt mound once or twice and getting a cut or two; certainly you didn't learn it from a video game.

But, maybe there is hope for the coming generations. At a car show a few years ago, a boy of ten or so was admiring my MG TD and finally came up to me. "Is that your car, mister?" "It's really neat." "I love the wood dash board." "Is that thing on the floor what they call a 'stick shift'?" Their numbers may be small, but I think there's hope for some of them.

Home Made Maple Syrup

This last item has nothing whatsoever to do with cars. I wrote it years ago and I don't recall for what. Thought I'd throw it in.

One of the fun things we used to do every spring was tap our maple tree in the back yard and make our own home-made maple syrup. I'll tell you right from the start that it is a bit of work and we don't get too much syrup, but it was fun.

You really only need to purchase one item. This is the little cast metal tube called a spile that you need to pound into the tree. We bought our spile some years ago at a local nature center. I've also seen some on e-Bay in the $5.00 price range. Everything else can be from home, as I'll explain.

To summarize the process, you drill a hole in the tree, tap the spile into the hole, attach a container to catch the sap, then boil the sap to make the syrup.

What type of tree? While a sugar maple tree produces the sweetest sap, and therefore needs to be boiled the least, any maple tree will work. You just have to boil a lot more sap to make it into syrup. I have a silver maple, and it makes great-tasting syrup, just not a lot for a whole lot of sap collected.

When do you tap the tree? You need to tap the tree during late Winter or early Spring when the tree starts producing sap. The dates aren't important, it's the temperature fluctuations between day and night that matter. To get a good flow running through the tree, the temperature needs to drop into the 20s at night and rise into the 30s or 40s during the day. The flow will stop if the temperature doesn't fluctuate above and below freezing each day and night. On a good day, I've collected over a gallon of sap.

Tapping the tree. I drill a 3/8" hole about an inch deep, then gently pound the spile into the hole. You can buy a metal bucket with a hole drilled in the side to hook the bucket over the spile so the sap drips into the bucket. They even come with a lid to keep out debris. I use a one-gallon milk jug tied to the spile with a piece of wire. I slip a short piece of clear plastic tubing over the spile to ensure all the sap goes into the jug. Now, sit back and watch the jug fill. On a sunny day, when it was below freezing the night before, you'll see sap drip at about one drop per second. You can save the sap in several milk jugs, or pour it into another container.

Cooking the sap into syrup. I boil the sap on the kitchen stove over a low fire. Many people prefer to cook the sap on a wood fire, which they claim gives it a better taste, but I keep it simple. As the gallon jug fills, I pour it into a large cooking pot. Rather than wait until I have a large quantity, I cook it a little bit at a time. I try to do the cooking on cold nights. The stove produces heat, relieving the furnace a little. Also, a gallon or so of moisture added to the air improves humidity, but cooking too much in one day may result in condensation on your windows. When I collect another gallon of sap, I add it to the pot and boil it again. If you cook the sap in the evening as I do, don't forget to shut off the fire before going to bed, or risk waking up in the middle of the night to a sweet burning smell and a lost batch of syrup (don't ask how I know). As it cooks, lots of sediment will accumulate in the pot. After you've cooked out about 80% of the water, the syrup still will be watery. At this point, you need to filter it. If you wait until the syrup is thick, it won't flow through the filter. If you don't have a large funnel, just use the top of a milk jug as I do, and a piece of paper towel to filter out the sediment. Then, cook the syrup again to its final thickness.

Then, make your favorite type of pancake or waffle, and enjoy!

Exotic equipment that I use. You'll need to find a spile and a short piece of tubing.

The spile inserted in the tree.

The tubing slid over the spile.

The milk jug is hung from the spile using a piece of string or wire.

Boiling the sap.

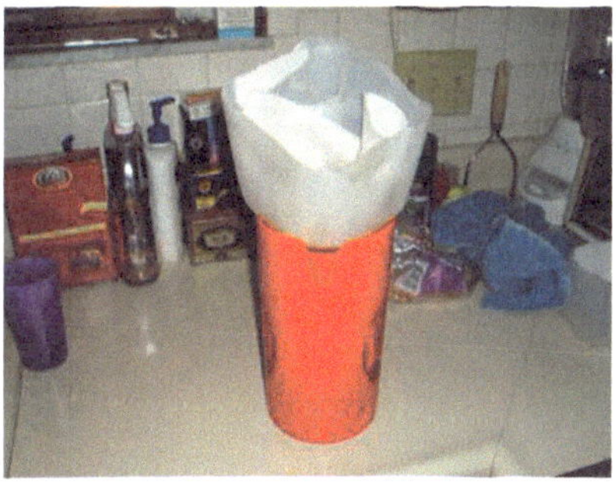
My exotic filtering system using the top of a gallon milk jug and a piece of paper towel.

British Stuff

Honoring a Special Veteran

Elizabeth wears an officer's uniform and stands beside an Auxiliary Territorial Service first aid truck during World War II.

The British monarchy has a long tradition of military service, but there has only been one woman from the British royal family ever to serve in the Armed Forces, and she's the only remaining head of state (as of 2022) to have served in World War II. If you haven't figured it out by now, I'm talking about Queen Elizabeth II.

When WWII raged in Europe, nearly everyone stood up to defend their homeland. Men, women, farmers, and businessmen did their duty alike. This includes then-Princess Elizabeth. Like her father, who served in WWI, she enlisted on her 18th birthday despite being in the line of succession for the throne and in spite of her father's reluctance.

Wearing a pair of coveralls, Princess Elizabeth trained in London as a mechanic and military truck driver.

After months of begging her father to let his heir pitch in, Elizabeth joined the Women's Auxiliary Territorial Service (ATS) as a Second Subaltern (from 1941 to 1949, the ATS used the ranks of second subaltern and subaltern, which were equivalent to second lieutenant and lieutenant respectively). The King ruled after long deliberations with his councilors that her training as a princess outweighed the nation's increasing manpower problems and that 'Betts' should not join any of the women's auxiliaries, nor work in a factory. But Betts had other ideas. It was not surprising that not long afterward the Palace made a straight-face announcement that the King 'had been pleased to grant an honorary commission as second subaltern in the ATS to Her Royal Highness the Princess Elizabeth.'

Princess Elizabeth enrolled in the Women's Auxiliary Territorial Service (ATS), similar to the American Women's Army Corps, where many women actively served in highly valuable support roles. Responsibilities of the ATS included serving as radio operators, anti-aircraft gunners, spotlight operators, and, the queen-to-be's specialty, mechanic and driver.

Once in the Auxiliary Territorial Service, Elizabeth learned how to change a wheel, deconstruct and rebuild engines, and drive ambulances and other vehicles. It wasn't a lavish position, but despite the grit and grime, she didn't just symbolically change a single tire and call herself a mechanic. She took her duties very seriously, taking great pride in her work. *Collier's Magazine* wrote at the time that "one of her major joys was to get dirt under her nails and grease stains on her hands."

However, being a princess still had its perks. She returned to the palace each night to sleep. On VE Day, then Princess Elizabeth slipped away with her sister to cheer with the crowds. The war was finally over and no one recognized the Princesses as they walked through the crowd incognito.

Less than a decade later, Princess Elizabeth would be crowned the Queen of England. Her independent spirit has endured to this day, as she never liked being chauffeured around when she could drive herself.

British Number Plates

(Editor's note – We call them license plates here in the States)

Whenever I saw a British license plate, I always wondered if there was any sort of pattern to what appeared to be a random mix of numbers and letters. Guess what…it turns out there actually is a system to it. With the help of the internet, the depository of all human knowledge, here is what I found out. BTW – they changed the numbering system so many times as the population of cars grew and grew that I'm not sure I got it exactly right, but here's my best guess.

Vehicle registration plates, referred to as 'Number Plates,' have been used in the United Kingdom since 1904. They are compulsory for motor vehicles used on public roads, with the exception of vehicles of the reigning monarch used on official business (that's because they're issued in the name of the Queen, so she doesn't need to issue one on behalf of herself). Basically, the groupings of characters tell when and where the car was registered, and then a unique code for that vehicle. With few exceptions, the plates cannot be transferred between vehicles.

The Motor Car Act of 1903 required all motor vehicles to be entered on an official vehicle register and to display the alphanumeric plates. The Act was passed so that vehicles could be traced easily in the event of an accident or violation of the law. Vehicle plates in the UK are rectangular or square in shape, with the dimensions of the plate and its lettering set down in law. Front plates are white, whereas back plates are yellow.

Within the UK there are two systems: one for Great Britain, which dates from 2001, and another for Northern Ireland, which is similar to the original 1903 system. Vehicles registered under previous numbering systems continue to retain their original number plates even though the area identifier in the previous number system is not the same as the post 2001 area identifier. Since I think all of our MGs were manufactured well before 2000, I'll limit my discussion to the older systems.

Plates issued between 1903 and 1932

This first system consisted of a one- or two-letter code followed by a sequence number from 1 to 9999. The code indicated the locality where the vehicle was registered. In England and Wales, these were initially allocated in order of population size (by the 1901 census) – thus A indicated London, B indicated Lancashire, C indicated the West Riding of Yorkshire and so on up to Y indicating Somerset, then AA indicated Hampshire, AB indicated Worcestershire and so on up to FP indicating Rutland.

The letters G, S and V were initially reserved for Scotland, and the letters I and Z for Ireland. In both cases, allocations of codes were made in alphabetical order of counties, followed by county boroughs – thus in Scotland, Aberdeenshire was allocated SA, Argyll received SB and so on, while in Ireland, Antrim was allocated IA, Armagh received IB, and so on.

When a licensing authority reached 9999, it became a bit harder to follow. The locality was allocated another two-letter code, but there was no pattern to these subsequent allocations as they were assigned on a first come first served basis. London and Middlesex quickly took most codes with L and M as the first letter respectively, while Surrey, initially allocated P, took many codes beginning with that letter.

The first mark to be issued in London was the simple, bold, A 1and this was registered to Earl Russell. He wanted the mark so badly he camped out all night to secure it..

A zero has been issued by several issuing authorities for the official car of the council head, in cases where plate number "1" had already been issued by the time the councils decided to give priority to its first citizen. Examples include the Lord Mayor of London (LM 0) and the Lord Provosts of Edinburgh (S 0), of Glasgow (G 0) and of Aberdeen (RG 0).

So much for the ancient history part.

1932 to 1963

Like I said earlier, the system initially made sense, but quickly became confusing as the popularity of motor vehicles grew. By 1932, the available codes were running out, and an extended scheme was introduced. This scheme placed an additional letter before the numerical code, but had the sequence number run only to 999, thus restricting the number of characters in a registration to only six. The first area to issue such plates was Staffordshire in July 1932 with ARF 1 etc. All other areas in England and Wales, plus most areas in Scotland, followed suit once they had issued all their two-letter registrations.

In some areas, the available alphanumeric combinations again started to run out in the 1950s, so they started what became known as "reversed" registrations – the letters coming after the numbers. Staffordshire was again the first area to issue such registrations, starting with 1000 E in 1953. In most cases, the three-letter combinations (e.g. 1 AXH for Middlesex) would be issued first, while in later years some areas started with the one- and two-letter combinations and others issued all three at the same time. The ever-increasing popularity of the car meant that by the beginning of the 1960s, these registrations were also running out. So much, again, for keeping it simple.

Some three-letter combinations were not used as they were deemed offensive. These included ARS, BUM GOD, SEX, and SOD. DUW was issued in London for several months in 1934 before it was realized it was the Welsh for "god", and withdrawn.

1963 to 1982

In August 1962, another attempt was made to create a national scheme again to alleviate the problem of registrations running out. This used the scheme introduced in 1932, of a three-letter combination followed by a sequence number from 1 to 999, but also added a letter suffix, which initially changed on January 1 each year. An "A" suffix thus was used for 1963, "B" for 1964, etc. Middlesex was the first authority to adopt this scheme when it issued AHX 1A in February 1963. Most other areas followed suit during 1964, but some chose to stick to their own schemes up until January 1965, when the letter suffix was made compulsory.

As well as yielding many more available numbers, this system was a handy way to identify the car's age. However, the year letter changing on January 1 each year meant that car retailers soon started to notice that buyers would tend to wait until the New Year for the new letter to be issued, so it would look like they were getting a "newer" car. This led to major peaks and troughs in sales over the year, so the industry lobbied to get the scheme changed to help level out sales. So, in 1967, the change of year letter was started on August 1 rather than January 1. Looks like the confusion is starting again.

In the beginning, plates were black with white or silver characters. Reflective plates were introduced in 1969 and were white on the front and yellow with black characters on the rear.

In October 1974, responsibility for issuing registrations was transferred from local and regional authorities to specialist Local Vehicle Licensing Offices (LVLOs) or Vehicle Registration Offices (VROs). Most of the two-letter area codes allocated during the first scheme continued in their respective areas, albeit now indicating the nearest LVLO/VRO rather than the local or regional authority. However, the decision to streamline the allocations of these codes meant that some were transferred to new areas. For instance, the former Suffolk code CF was transferred to Reading, while the former Edinburgh code WS was re-allocated to Bristol.

1983 to 2001

By 1982 suffixes had reached Y, and so from 1983 onwards the sequence was reversed again, i.e., that the year letter — starting again at "A" — preceded the numbers. The available range was then A21 AAA to Y999 YYY, the numbers 1–20 being reserved for the government, and DVLA select registration sales scheme. Towards the mid-1990s there was some discussion about introducing a unified scheme for Europe, which would also incorporate the country code, but after much debate such a scheme was not adopted because of lack of countries willing to participate. (Thank goodness!)

The changes in 1983 also brought the letter Q into use – although on a very small and limited scale. It was used on vehicles of indeterminate age, such as those assembled from kits, substantial rebuilds, or imported vehicles where the documentation would be insufficient to determine the age. There was a marked increase in the use of Q registrations in the late 1980s and early 1990s, fuelled by car crime. Many stolen vehicles had false identities given to them, and when this was discovered and the original identity could not be determined, a Q registration would be issued to such vehicle. It was seen as an aid to consumer protection. Due to indeterminate age, origin and specification of Q registration vehicles, most motor insurers are reluctant to insure these 'Q-plate' vehicles.

By the late 1990s, the range of available numbers was once again starting to run out, but I'll cut this off at this point. As Dr. Suess said in one of his books of tongue twisters, "Is your tongue numb yet?" If you have a British number plate on your MG, and the above discussion helped you to sort out where the plate came from and what it meant, my hats off to you. Please don't ask me to repeat any of the above from memory.

History of the 'British' Flag

As British car enthusiasts, we all have some connection to the British flag, be it in our club logo, stickers on our cars, or the design on t-shirts. But, did you ever wonder where the design came from? It is an intricate design, so you probably suspect that there's a story and some history to it (after all, it *is* British), and you're right. So, here's an abbreviated version of the history of the British flag.

First national flag of England, the Cross of St. George

First, you need to understand the difference between England, Great Britain, and the United Kingdom. They are not interchangeable, although we often use them that way. First, Great Britain is an island. Second, England is one of several states on the island of Great Britain. And last, the United Kingdom is a country made up of the member states of England, Scotland, Northern Ireland, and Wales. So, it's really not even a British flag, but the flag of the United Kingdom. But, we'll overlook that for the sake of this discussion. Let's go back about 800 years ago. At this point in the history of the United Kingdom, England, Scotland, Wales and Ireland were separate countries. However, this was soon to change. In 1194 A.D., Richard I of England introduced the Cross of St. George, a red cross on a white background, as the National Flag of England.

In 1536, under Henry VIII, the Act of Union was passed making Wales, in effect a province of England (it never was conquered, but joined "peacefully," and the flag of England became the flag of Wales as well).

After Queen Elizabeth I of England died in 1603, King James VI of Scotland inherited the English throne and became King James I of England. But, the two countries were not united. Each country kept its own parliament and flag.

Scotland's Flag of St. Andrew, a blue field with a white diagonal cross, called a 'saltire.'

Early in his reign James attempted to combine England and Scotland in a united kingdom of 'Great Britain'. He presented this to his first Parliament, on 22 March 1604. The union was resisted, but James defied them. On 20 October 1604 he proclaimed a new title for himself as King of Great Britain, but essentially he was the king of two separate countries. A problem arose, which flag should be hoisted on the king's ships; English sailors resented the Scottish colors and the Scots scorned the cross of St. George.

The problem was solved in 1606 when a compromise led to the creation of the first Union Flag. On 12 April 1606, the national flags of Scotland and England were united into a single flag, the first 'Union Jack'. However, this flag was used only at sea for over a century. Ashore, the old flags of England and Scotland continued to be used by their respective countries. Ever wonder where the term 'Union Jack' originated? The forward-most mast on sailing ships of the time was the 'jack' mast, and since the national flag was flown on the jack mast and only at sea, it became known as the 'Union Jack.'

The first 'Union Jack.' Heraldry rules demanded that two colors must never touch each other, so a white border was added around the red cross.

It wasn't until 28 July 1707, during the reign of Queen Anne, that a royal proclamation made the Union Jack the national flag of Great Britain, for use ashore and afloat. The Act of Union of 1707, during the reign of Queen Anne, formally joined England and Scotland together, finally creating a single kingdom with a single Parliament, called the 'United Kingdom of Great Britain'. This act accomplished what King James I had failed to do a century earlier. England, Wales and Scotland now were united together under one monarch and one parliament. Since the national flag now was used on land also, technically the term 'Union Jack' became obsolete, but remains in common use.

On 1 January 1801, Ireland joined Great Britain and it became necessary to represent Ireland in a new national flag. This union resulted in the flag that has been flown ever since.

The Cross of St. Patrick, Ireland's national flag.

The appropriate placement of the white and red crosses (saltires) must have presented some challenges, but eventually it was worked out. The symbols of Scotland and Ireland (the white and red saltires) were placed side by side on the Union Flag. But, since Scotland joined the Union nearly two hundred years before Ireland, the white St Andrew's Cross was placed uppermost in the top left quarter, this being the most honorable position according to heraldry, while the red Irish Cross was given the position below it. But, notice that on the opposite quarter, the red is on top of the white! In heraldry, this is known as "counterchanging." Thus, the new British flag is not symmetrical. Last, in order to avoid having the red of the Irish Cross directly touch the blue field of the Scottish Cross, an edging of white was added to the red Irish Cross. So, while giving the appearance of a red stripe that is offset on a white background, it actually is the white and red saltires placed side-by-side, plus a white border, arranged according to strict rules of heraldry.

The UK's current flag, in use since 1801. This is how it should look when the staff is on the left. It is not uncommon to see the flag accidentally flown upside down!

England, Wales, Scotland and Ireland were now all joined together and called the United Kingdom of Great Britain and Ireland. The name was later changed to United Kingdom of Great Britain and Northern Ireland when the greater part of Ireland left the United Kingdom in 1921. This is its correct name today.

Thus, the 'British' flag, which, to be correct, really should be called the flag of the United Kingdom of Great Britain and Northern Ireland, is a merging of the crosses of St George, St Andrew and St Patrick, the Patron Saints of England, Scotland and Ireland. But, why is Wales not represented on the Union Flag? What ever happened to Wales?

The flag of Wales

The Welsh dragon does not appear on the flag because when the first union flag was created in 1606, Wales was already united with England from the 13th century. This meant that Wales was a principality instead of a kingdom and as such could not be included. However, with the rising trend of nationalism and political correctness, there is a movement to have the Welsh dragon incorporated into the "British" flag. I'll leave it to you figure out how heraldry and counterchanging rules would apply!

So, now you know.

Editor's note - I lifted this information from numerous websites, but mostly from: http://www.woodlands-junior.kent.sch.uk/geography/unionjack.html

What/Who is Aunt Sally?

One of the fun things we do at club picnics (actually, everything is fun) is playing a game in which we attempt to knock a wood doll's head from a post by throwing sticks at it. It's called Aunt Sally. I always wondered about the game's origins, so again went to the source of all information, the internet. Here's what I learned.

Aunt Sally is a traditional English game usually played in pub gardens and fairgrounds, and dates back to the 17th Century. Leagues of pub teams, each consisting of eight players, still play the game today, throughout the spring and summer months, mainly in Oxfordshire and some bordering counties.

One theory is that in Port Meadow in Oxfordshire, at the time of the English Civil War, King Charles I set up court in Oxford. The Cavaliers (soldiers loyal to King Charles I) were bored and formed a game with sticks and makeshift materials similar to the game we play today. An Aunt Sally was originally the modeled head of an old woman with a clay pipe in her mouth, the object being for players to throw sticks at the head in order to break the pipe. It later was simplified to a ball on a stick.

There are other theories, but by Victorian times the game became somewhat of a crude fairground game. The doll was dressed up to resemble an old maid and no doubt it was thought to be amusing to chuck sticks at the ugly looking old woman. Sometimes the figure was painted black, thus making the game both racist and anti-women. It has been suggested that the name “Aunt Sally” was based on the blackface doll, inspired by a low-life character named "Black Sal", which appeared in an 1820s series of novellas titled *Life of London* by Pierce Egan, a contemporary of Charles Dickens. Aunt Sally also has become another name for a debating tactic in which someone purposefully misrepresents an argument with the express intent of inviting criticism, or to encourage an idea to be discussed and refuted.

The game itself is simple enough. The head rests on a short post, about 2 ½ feet above the ground. Players throw sticks at the head, using an underhand toss, trying to knock the head off without hitting the post. Successfully hitting the dolly off is known as a "doll"; however if the spike is hit first, then the score does not count and is called an "iron". Normal league play has two teams each consisting of eight players and three legs or "horses" are played. Each horse consists of each member of each team having one turn so that each team makes 48 throws. It is believed that the record for a horse stands at 40 so it can be seen that hitting the doll is quite tricky.

Besides being played at our club picnics, as mentioned above, Aunt Sally matches still can be found in England. In 2011 the inaugural Aunt Sally Singles World Championship took place at the Charlbury Beer Festival in Charlbury, West Oxfordshire which was attended by former prime minister David Cameron and the tournament has continued there annually ever since.

In Memory

Remembering some friends we've lost along the way with whom we shared good times and many happy miles.

Gary Kinney (2007)

Jasper Cemino (2009)

John Zeno (2010)

Barbera Smith (2011)

Dick Goodman (2012)

Joe Hooker (2013)

Bill Hammond (2014)

Pat & Reuben Wasserman (2014, 2019)

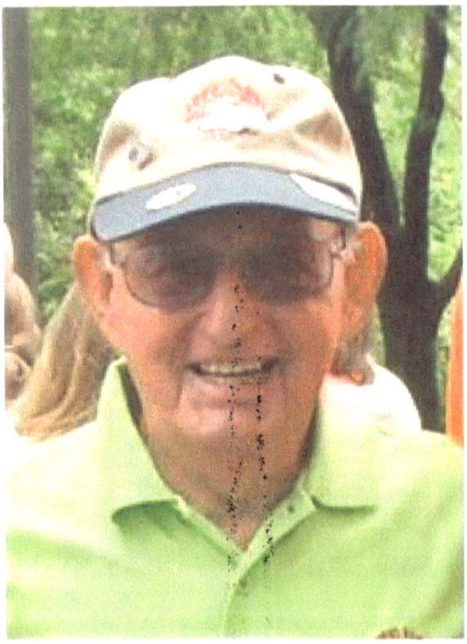

Louie DiPasquale (2015)

Bob Kraft (2017)

Dick Smith (2019)

Phil Johnson (2020)

Bob Farrell (2021)

Jim Carson (2021)

Meet Steve Markman

Steve's not really a car guy, but he loves his MG TD, probably the best known classic British roadster. As editor of his car club's newsletter, and often needing a few additional pages of material, he started writing articles on various automotive-related topics. Things like explaining octane, lubricants, paint, and even how aviation technologies found their way into your car. These soon grew into three or four pieces each year. This book is a compilation of many of his articles.

Steve was an Air Force maintenance officer for Minuteman nuclear missiles, and then worked as a civilian aerospace engineer, first developing engineering flight simulators, than managing flight research aircraft. He even earned commercial, instrument, and multi-engine pilot ratings. Steve also is the author of two books on research aircraft, and two personal memoirs.

Besides aviation and his MG TD, Steve's other passions are working with wood, building sailing ship models, keeping his koi pond and tropical fish, martial arts, writing, home maintenance, and just about anything else that comes up.

Made in United States
Troutdale, OR
11/03/2026

56845391R00106